T0094266

BIOLOGICAL AND MEDICAL PHYSICS, BIOMEDICAL ENGINEERING

For other titles published in this series, go to
www.springer.com/series/3740

BIOLOGICAL AND MEDICAL PHYSICS, BIOMEDICAL ENGINEERING

The fields of biological and medical physics and biomedical engineering are broad, multidisciplinary and dynamic. They lie at the crossroads of frontier research in physics, biology, chemistry, and medicine. The Biological and Medical Physics, Biomedical Engineering Series is intended to be comprehensive, covering a broad range of topics important to the study of the physical, chemical and biological sciences. Its goal is to provide scientists and engineers with textbooks, monographs, and reference works to address the growing need for information.

Books in the series emphasize established and emergent areas of science including molecular, membrane, and mathematical biophysics; photosynthetic energy harvesting and conversion; information processing; physical principles of genetics; sensory communications; automata networks, neural networks, and cellular automata. Equally important will be coverage of applied aspects of biological and medical physics and biomedical engineering such as molecular electronic components and devices, biosensors, medicine, imaging, physical principles of renewable energy production, advanced prostheses, and environmental control and engineering.

Victor Bloomfield

Computer Simulation and Data Analysis in Molecular Biology and Biophysics

An Introduction Using R

 Springer

Victor Bloomfield
Department of Biochemistry, Molecular
  Biology, and Biophysics
University of Minnesota
321 Church Street
6-155 Jackson Hall
Minneapolis, MN 55455
USA
victor@umn.edu

ISSN 1618-7210
ISBN 978-1-4419-0084-5   e-ISBN 978-1-4419-0083-8
DOI 10.1007/978-1-4419-0083-8
Springer Dordrecht Heidelberg London New York

Library of Congress Control Number: 2009926315

Printed on acid-free paper

springer.com

Preface

This book provides an introduction to two important aspects of modern biochemistry, molecular biology, and biophysics: computer simulation and data analysis. My aim is to introduce the tools that will enable students to learn and use some fundamental methods to construct quantitative models of biological mechanisms, both deterministic and with some elements of randomness; to learn how concepts of probability can help to understand important features of DNA sequences; and to apply a useful set of statistical methods to analysis of experimental data. The availability of very capable but inexpensive personal computers and software makes it possible to do such work at a much higher level, but in a much easier way, than ever before.

The Executive Summary of the influential 2003 report from the National Academy of Sciences, "BIO 2010: Transforming Undergraduate Education for Future Research Biologists" [12], begins

> The interplay of the recombinant DNA, instrumentation, and digital revolutions has profoundly transformed biological research. The confluence of these three innovations has led to important discoveries, such as the mapping of the human genome. How biologists design, perform, and analyze experiments is changing swiftly. Biological concepts and models are becoming more quantitative, and biological research has become critically dependent on concepts and methods drawn from other scientific disciplines. The connections between the biological sciences and the physical sciences, mathematics, and computer science are rapidly becoming deeper and more extensive.

Quantitative approaches have become particularly prominent in the large-scale approaches of systems biology and its associated high-throughput techniques: bioinformatics, genomics, proteomics, metabolomics, cellomics, etc. High levels of quantitation are also needed in some of the more biophysically oriented aspects of biochemistry, molecular and cellular biology, physiology, pharmacology, and neuroscience.

The increasing use of quantitation at the frontiers of modern biology requires that students learn some basic quantitative methods at an early stage, so that they can build on them as their careers develop. To deal with realistic biological problems, these quantitative methods need to go beyond those taught in standard courses in calculus and the elements of differential equations and linear algebra—courses based mainly on analytical approaches—to encompass appropriate numerical and computational techniques. The types of realistic biological problems that contemporary science is facing are generally too large and complex to yield to analytical approaches, and usually specific numerical answers are desired, so it makes sense to go directly to computational rather than analytical mathematical answers.

Modern molecular and cellular biology also demands increasingly sophisticated use of statistics, a demand difficult to meet when many life science students don't take even an elementary statistics course.

To add significant instruction in computational and statistical methods to an already overcrowded biology curriculum poses a challenge. Fortunately, modern computer tools, running on ordinary personal computers, enable very sophisticated analyses without requiring much analytical or programming knowledge or effort. In essence, this is a "black box" approach to quantitative biology; but I would argue that using a set of black boxes is better than not using quantitative tools at all when they would substantially enhance the results of biological investigations. The challenge, then, is to make students—and more mature scientists—aware of the appropriate black boxes, their capabilities, and the steps needed to access those capabilities. This will require a small amount of programming and an even smaller amount of analytical manipulation.

In this book I show how to use readily available computer tools to formulate quantitative models and analyze experiments in a way that measures up to the standards of biology in the 21st century. In particular, I show how to use the free, open-source software program R in a variety of biological applications.

R is a free software environment for computer programming, statistical computing, and graphics. The R web site emphasizes statistical computing and graphics, which it does superlatively well; but R is also a very capable environment for general numerical computer programming.

The characteristics of R that make it a good choice on which to build quantitative expertise in the biochemical sciences are:

- It runs on Mac OS, Windows, and various Linux and Unix platforms.
- It is free, open-source, and undergoing continual (but not excessive) development and maintenance. It is an evolving but stable platform that you will be able to rely on for many years.
- It has a wide variety of useful built-in functions and packages, and can be readily extended with standard programming techniques.
- It has excellent graphics.
- Its capabilities are very similar to excellent and widely-used but expensive commercial programs such as Matlab.
- If needed for large, computationally demanding projects, R can be used to interface with other, speedier but less convenient programming languages.
- Once you learn its (fairly simple) syntax, it is easier and more efficient than a spreadsheet.

- It has many sample datasets, which help with learning to use the program.
- It is widely used in statistics, and is increasingly used in biological applications. See particularly the Bioconductor project primarily based at the Fred Hutchinson Cancer Research Center. Bioconductor is "an open source and open development software project for the analysis and comprehension of genomic data."

Because of these capabilities, R can serve as your basic quantitative, statistical, and graphics tool as you develop your career.

Useful references

The book that comes closest to this one in emphasizing the numerical and programming capabilities of R, rather than mainly its statistical capabilities, is *Introduction to Scientific Programming and Simulation Using R* by Owen Jones, Robert Maillardet, and Andrew Robinson, Chapman & Hall/CRC (2009) [36]. It has particular emphasis on probability and stochastic simulation.

Most of the books that teach how to use R (or its progenitor S) do so in the context of its use as a program for doing statistics. Statistics will be only one of our foci in this book, but one or more of these books may be useful for reference.

- *Introductory Statistics with R*, by Peter Dalgaard, Springer (2002) [15]
- *Using R for Introductory Statistics*, by John Verzani, Chapman & Hall/CRC (2005) [65]
- *Modern Applied Statistics with S*, 4th ed. by W.N. Venables and B.D. Ripley, Springer (2002). (The standard advanced reference.) [64]
- *Data Analysis and Graphics Using R: An Example-Based Approach* (Cambridge Series in Statistical and Probabilistic Mathematics), 2nd ed., by John Maindonald and John Braun, Cambridge University Press (2007) [43]

Several books use R in a biological context. An excellent online text at a level similar to this book is K. Seefeld and E. Linder. *Statistics Using R with Biological Examples* [55].

An advanced introduction is *Computational Genome Analysis: An Introduction* (Statistics for Biology & Health) by Richard C. Deonier, Simon Tavaré, and Michael S. Waterman, Springer (2005) [18]. The emphasis of this book is mostly on bioinformatics.

R Programming for Bioinformatics by Robert Gentleman, Chapman & Hall/CRC (2008) [26] deals with programming in R, with bioinformatics examples. It assumes a good deal of prior programming experience.

Analysis of Phylogenetics and Evolution with R by Emmanuel Paradis, Springer (2006) [49] has a useful introduction and some intermediate to research-level examples in both DNA and organismal phylogenetics and evolution.

Stochastic Modelling for Systems Biology by Darren J. Wilkinson, Chapman & Hall/CRC (2006) [67] uses some R code in its treatment of systems biology.

Bioinformatics and Computational Biology Solutions Using R and Bioconductor (Statistics for Biology and Health) edited by Robert Gentleman, Vincent Carey, Wolfgang Huber, Rafael Irizarry, and Sandrine Dudoit, Springer (2005) [27] explores the Bioconductor packages, especially as applied to analysis of microarray data. A compact book that is perhaps more suitable as an introduction to this material is *Bioconductor Case Studies* edited by Florian Hahne, Wolfgang Huber, Robert Gentleman, and Seth Falcon, Springer 2008 [31].

In my development of this book and course, I have drawn heavily on *Computer Simulation in Biology: A BASIC Introduction*, by R.E. Keen and J.D. Spain (1992) [39]. This book, which appears to be out of print, uses BASIC, an earlier and less capable computer language than R; but it has a good selection of topics and computer simulation examples for an introductory course.

Of the many recent books on mathematical and computational biology, these two fall closest to my philosophy, in their selection of topics and in emphasizing computational rather than analytical approaches:

- *Computational Cell Biology* edited by Christopher Fall, Eric Marland, John Wagner, and John Tyson, Springer (2002) [24]
- *Mathematical Models in Biology: An Introduction*, Elizabeth S. Allman and John A. Rhodes, Cambridge University Press (2004) [2]

Acknowledgments

I thank Professors David Bernlohr and Paul Siliciano for having arranged the opportunity to teach the course, Biochemistry 4950, that led to this book. I am grateful to the students in the course for giving feedback about the material and pointing out puzzlements and inconsistencies. I am particularly grateful to Geteria Onsongo for noting many typos.

Contents

Part I
The Basics of R

Chapter 1
Calculating with R

In this chapter, we begin with the most basic aspects of R: installing it, checking the installation by demonstrating some of its impressive graphics capabilities, and showing how it can be used as a powerful calculator. We will go well beyond these capabilities in subsequent chapters, but they should demonstrate that R can be a constant presence on your desktop for the kinds of calculations and graphs for which you would otherwise use a handheld graphics calculator or spreadsheet.

1.1 Installing R

R can be downloaded and installed from the CRAN (Comprehensive R Archive Network) web site at http://cran.r-project.org/. Click on the link for your operating system, and follow the directions. You should choose the pre-compiled binary rather than the source code.

1.1.1 Demos

Once you've installed and launched R, you will probably want an immediate demonstration of some of its capabilities. R's graphics capabilities are excellent and very flexible. To see an illustration of the kinds of graphics that R is capable of, type `demo(graphics)` at the prompt (>), and press Return to view the successive graphs. To see further demonstrations of contour plots, type `demo(image)` at the prompt.

V. Bloomfield, *Computer Simulation and Data Analysis in Molecular Biology and Biophysics,*
Biological and Medical Physics, Biomedical Engineering, DOI: 10.1007/978-1-4419-0083-8_1,
© Springer Science + Business Media, LLC 2009

1.2 Finding help with R

The R program itself, as it is installed on your computer, has a very extensive Help facility. Look for it in the menu. In particular, "An Introduction to R" and "R Data Import/Export" are likely to be useful as you begin learning the language. These and other manuals are available both through the R Help menu and the CRAN web site (Documentation/Manuals).

At any time while you're in the R console or command line, you can look for help on a particular topic by typing `help(topic)` or `?topic`. Typing `?help` will give you information about the help system itself. A problem with the R help system is that you generally have to know the exact term being searched for, since the help system searches a pre-established index rather than the full text. For example, trying to learn about correlation analysis by typing `help(correlation)` or `?correlation` yields "Help topic not found". `?corr` gives the same result. Finally, `?cor` brings up the desired help page. The functions `apropos` and `help.search` may (or may not) be useful in such cases. Two aids to finding online help about R topics are *RSeek* `http://www.rseek.org/` and *Search the R Statistical Language* `http://www.dangoldstein.com/search_r.html`.

There are numerous online sites devoted to R. A particularly useful one for rapid reference is R & BioConductor Manual by T. Girke at UC Riverside.
`http://faculty.ucr.edu/~tgirke/Documents/`
`   R_BioCond/R_BioCondManual.html`
Girke writes:

> This manual provides a condensed introduction into the basic R and BioConductor functions. A not always very attractive, but practical copy&paste format has been chosen throughout this manual. In this format all commands are represented in bold font followed by a short description of their usage in non-bold font. Often several commands are concatenated on one line and separated with a semicolon ';'. All explanations start with the standard comment sign '#' to prevent them from being interpreted by R as commands.

These short examples are very handy to learn how to use the various commands and to inspect sample output.

Another very handy resource is the online "R Reference Card" by Tom Short:
`http://cran.r-project.org/doc/contrib/Short-refcard.pdf`

The R Wiki (`http://wiki.r-project.org/rwiki/doku.php`) "is dedicated to the collaborative writing of R documentation." If you are stymied by the sometimes overly terse official R documentation, this site may provide additional helpful information.

1.3 Some interface aids

Your first steps in using R will largely involve one-line commands. When you begin to write more involved, multi-line programs, you will find it advantageous to use a text editor (either the one in R or one that comes with your computer, such as TextEdit (Mac), Notepad (Windows), or vi (Unix)) to write and edit the code. You can then copy and paste the code into R and run it by hitting Return. Typos and other mistakes are much easier to correct in a text editor than from the R console.

If you want to reuse, or correct, a bit of code used recently, the Up-Arrow key will scroll through previous commands. You can also retrieve previous commands with the Command History window.

1.4 Arithmetic

R has the standard numerical operations:

```
> 2+2
[1]  4
> # Two commands on same line, separated by semicolon
> 2*3; 2/3
[1]  6
[1]  0.6666667
> # Default is seven digits after the decimal.
> # To change:
> options(digits=4)
> 2/3
[1]  0.6667
> 2.1^3
[1]  9.261
> 17%/%8 # integer division
[1]  2
> 2-3
[1]  -1
```

The leading > is a prompt. It appears automatically, you don't need to type it. Everything after the # is treated as a comment. The [1] indicates the position of the first response on that line. When there is only a single response, the [1] seems unnecessary; but as you'll soon see, when your answer involves a sequence of 100 numbers, with line breaks every six or eight, the index at the beginning of each line provides helpful orientation.

1.5 Complex numbers

R knows about imaginary and complex numbers, with i as sqrt(-1):

```
> (3+2i) + (4-7i)
[1]  7-5i
```

```
> (3+2i) * (4-7i)
[1] 26-13i
> (3+2i) / (4-7i)
[1] -0.0308+0.4462i
```

To work with complex numbers, supply an explicit complex part. Thus sqrt(-17) will give NaN and a warning, but sqrt(-17+0i) will do the computations as complex numbers.

1.6 Assigning variables

R is case sensitive, so X and x are different symbols and refer to different variables. All alphanumeric symbols (a–z, A–Z, 0–9) are allowed plus "." and "_", with the restriction that a name must start with "." or a letter, and if it starts with "." the second character must not be a digit.

You can assign numerical values to symbolic variables using either $=$ or $< -$. The first is more familiar and accords better with usage in other computer languages; the second was the original way in R, and perhaps better symbolizes the concept of assignment. Recent books on R most commonly use $=$, and I will follow that usage. As we'll see later, $=$ is also used in R to set values for options in functions and procedures. Both of these should be distinguished from $==$, which is the "logical equals" that we will discuss later.

```
> x = 10
> y = -4
> x/y
[1] -2.5
> x^y
[1] 1e-04
> pi # R knows the value of pi
[1] 3.142
```

1.6.1 Example: Conversions between units

As a simple but useful example of how to use named variables, we show how to convert from one system of units to another. (We'll adopt the convention of designating units by a terminating period.) Let's consider units of energy. In the SI system, the basic unit of energy is the joule (J.). In chemical work, it is common to use calories (cal.); and in nutrition, to use kilocalories (kcal.) which are often denoted Cal (Cal.) In physics, the erg (erg.) is frequently used. We can set up conversions between these units in R:

```
> J. = 1; cal. = 4.184*J.; kcal. = 1000*cal.
```

```
> Cal. = kcal.; erg. = 1.0e-7*J.
```

Then we can calculate the number of joules or ergs in a 280-Cal candy bar by

```
> options(digits=4)
# Note how "=" is used to set the value of an option.
> 280*Cal./J.
[1] 1171520
> 280*Cal./erg.
[1] 1.172e+13
```

In general, convert from unit.1 to unit.2 by multiplying by the ratio unit.1/unit.2. (The reason to divide is that your final answer is a pure number, not a quantity with units.)

1.7 Standard mathematical functions

R contains the standard functions, which work on numbers or on variables. For example:

```
> x = 2
> log(x,2)
[1] 1
> log(4,2)
[1] 2
> x = 1.7
> sin(x)
[1] 0.9917
> cos(x)
[1] -0.1288
> tan(x)
[1] -7.697
> atan(x)
[1] 1.039
> sinh(x)
[1] 2.646
> log(x) # Natural log, base e
[1] 0.5306
> log10(x) # Base 10 logarithm, also as log(x,10)
[1] 0.2304
> log(4,2) # Base 2
[1] 2
> exp(x)
[1] 5.474
> sqrt(x)
```

```
[1] 1.304
```

1.7.1 Rounding

R has a handy set of functions for rounding numbers according to various criteria.
See ?round in R help for details.

ceiling(x)   returns the smallest integer not less than x.
floor(x)   returns the smallest integer not greater than x.
trunc(x)   returns the integer toward 0.
round(x,d)   rounds x to the specified number of decimal places d, with default
    d = 0.
signif(x,d)   rounds x to the specified number of significant digits d, with default
    d = 6.
zapsmall(x,d)   treats very small values of x as 0.

In these examples, we anticipate the next section and apply these functions to vectors
of several numbers, in which case the functions act on each number individually.

```
> x = c(1.2345, -1.2345, 1.2345e-15)
> ceiling(x)
[1]   2 -1   1
> floor(x)
[1]   1 -2   0
> trunc(x)
[1]   1 -1   0
> round(x,3)
[1]   1.234 -1.234   0.000
> signif(x,3)
[1]   1.23e+00 -1.23e+00   1.23e-15
> zapsmall(x,3)
[1]   1.234 -1.234   0.000
```

1.8 Vectors

R operates on data structures. The simplest such structure is the numeric vector,
which consists of an ordered collection of numbers, separated by commas and
bracketed by c(), which stands for "concatenation":

```
> x = c(10.4, 5.6, 3.1, 6.4, 21.7)
```

A scalar, or single number, can be thought of as a vector with just one element.

1.8.1 Operations on vectors

Once a vector is defined, most of the numerical operations act on each element separately, which is very convenient.

```
> x = c(10.4, 5.6, 3.1, 6.4, 21.7)
> x+2
[1] 12.4  7.6  5.1  8.4 23.7
> 2*x
[1] 20.8 11.2  6.2 12.8 43.4
> x/2
[1]  5.20  2.80  1.55  3.20 10.85
> x^2
[1] 108.16  31.36   9.61  40.96 470.89
```

If a shorter vector y acts on a longer vector x, the result is a vector with the length of x; y is *recycled* to match x. Fractional recycling is allowed, but with a warning.

```
> x = c(1,2,3,4,5,6,7)
> y = c(1,2)
> x + y
[1] 2 4 4 6 6 8 8
Warning message:
In x + y : longer object length is not a multiple of
shorter object length
```

Individual elements of a vector can be picked out using square brackets:

```
> x[2]
[1] 5.6
> x[c(2,4)]
[1] 5.6 6.4
```

To change an individual element of a vector:

```
> x[2] = 6.7
> x
[1] 10.4  6.7  3.1  6.4 21.7
```

Since a scalar is a vector of unit length, a scalar can be expanded to a vector of length greater than one as follows:

```
> x = 1 # x begins as a scalar
> x[2] = 2 # Now x becomes a vector of length 2
> x
[1] 1 2
```

We will find this to be useful when we numerically solve differential or difference equations as vectors of magnitude vs. time, starting with initial conditions (scalars) for those magnitudes.

Two vectors can be added, multiplied, etc.

```
> a = c(1,2,3)
> b = c(3,5,7)
> a+b
[1]   4  7 10
> a*b
[1]   3 10 21
> a/b
[1] 0.3333 0.4000 0.4286
```

Note that * gives the element by element product. To get the dot product of two vectors, use %*%, which is also the operator for matrix multiplication (see later).

```
> a%*%b
       [,1]
[1,]    34
```

In this result, the indexes [1,] and [,1] denote the first row and first column of a 1×1 matrix (a single number). We'll see later the generalization of this notation to larger matrices.

The standard functions act individually on the elements of vectors:

```
> options(digits=4)
> x =  c(10.4, 5.6, 3.1, 6.4, 21.7)
> sin(x)
[1] -0.82783 -0.63127  0.04158  0.11655  0.28705
> log(x)
[1] 2.342 1.723 1.131 1.856 3.077
> exp(x)
[1] 3.286e+04 2.704e+02 2.220e+01 6.018e+02 2.656e+09
> sqrt(x)
[1] 3.225 2.366 1.761 2.530 4.658
```

1.8.2 Functions that operate on vectors

There are also functions that calculate the length, mean, standard deviation, min, max, etc., of a vector:

```
> x =  c(10.4, 5.6, 3.1, 6.4, 21.7)
> length(x)
[1] 5
> mean(x)
[1] 9.44
> sd(x) # Standard deviation
[1] 7.338
> var(x) # Variance = sd^2
[1] 53.85
```

```
> min(x)
[1] 3.1
> max(x)
[1] 21.7
> range(x)
[1]   3.1 21.7
> sum(x) # Sum of the elements of x
[1] 47.2
> prod(x) # Product of the elements of x
[1] 25074
```

1.8.3 Character vectors

The components of vectors can be letters as well as numbers. This is useful in representing DNA and protein sequences. For example:

```
DNAseq = c("A","T","G","C")
AmAcseq = c("Ile","Pro","Gly","Gln")
```

1.9 Generating sequences

We will often want to generate sequences, i.e., vectors, of numbers to serve as bases for simulations. Such sequences might be regular, for example a sequence of regularly spaced time points for modeling reaction kinetics or population growth; or random, for example a set of random integers between 1 and 4 to simulate a sequence of the four DNA bases.

1.9.1 Generating regular sequences

To generate a sequence of numbers separated by 1, use the colon notation:

```
> 1:10
 [1]  1  2  3  4  5  6  7  8  9 10
> 10:1 # Counting backward
 [1] 10  9  8  7  6  5  4  3  2  1
> 5:-5
 [1]  5  4  3  2  1  0 -1 -2 -3 -4 -5
```

The colon has a higher priority than other operations:

```
> n = 10
```

```
> 1:n-1
 [1] 0 1 2 3 4 5 6 7 8 9
> 1:(n-1)
 [1] 1 2 3 4 5 6 7 8 9
```

To increment the sequence by a more general value than 1, use seq.

```
> s = seq(from =-2,to = 2, by=.5)
> s
 [1] -2.0 -1.5 -1.0 -0.5  0.0  0.5  1.0  1.5  2.0
```

If the values of from, to, and by are in that order, the names can be omitted. s = seq(-2, 2, 0.5) gives the same result. Sequences of given lengths, starting points, and increments; or lengths, starting, and ending points can also be generated:

```
# 9 values starting at -2, incrementing by 0.5
> s1=seq(length=9, from=-2, by=.5)
> s1
 [1] -2.0 -1.5 -1.0 -0.5  0.0  0.5  1.0  1.5  2.0
# 12 values starting at 1 and ending at 6
> s2 = seq(from = 1, to = 6, length = 12)
> s2
 [1] 1.000 1.455 1.909 2.364 2.818 3.273 3.727 4.182
 4.636 5.091 5.545 6.000
```

To repeat values in a sequence, use the rep() function.

```
> y = 3
> rep(y,times=5)  # Or rep(y,5)
[1] 3 3 3 3 3
> z = c(1,2)
> rep(z,5)
 [1] 1 2 1 2 1 2 1 2 1 2
> rep(z,each=5)
 [1] 1 1 1 1 1 2 2 2 2 2
```

1.9.2 Generating sequences of random numbers

We will consider random variables drawn from several different probability distributions later in this book. However, it is useful here to have a brief introduction to the generation of sequences of uniformly or normally distributed random numbers.

A sequence of n random numbers uniformly distributed between min and 1max is generated by runif(n,min,max). If min and max are omitted, the defaults are 0 and 1.

```
> runif(5,1,3)
[1] 1.583 1.318 1.789 2.311 1.581
> runif(5)
[1] 0.57440 0.31678 0.12809 0.03409 0.08702
```

Similarly, a sequence of *n* random numbers distributed according to the normal distribution with mean = mu and standard deviation = sd is generated by rnorm(n, mu, sd). If mu and sd are omitted, the defaults are 0 and 1.

```
> rnorm(5,10,2)
[1] 10.537 11.672  9.899  5.900  8.344
> rnorm(5)
[1] -1.0949  2.4247  0.7083  2.9834 -0.1248
```

For example, to generate a sequence of five linearly increasing numbers with normally distributed error with standard deviation ±0.05:

```
> x = 1:5
> err = rnorm(5,0,0.05)
> x + err
[1] 0.9302 2.0432 2.9979 3.9991 4.9671
```

Alternatively, to generate the sequence with 5% relative error:

```
> x*(1+err)
[1] 0.9302 2.0865 2.9938 3.9964 4.8357
```

We will use this method frequently to generate simulated "noisy" data.

1.10 Logical vectors

The components of vectors can also be the logical values TRUE (or T) and FALSE (or F). As an example, we can test which components of a numerical vector are negative:

```
> v = rnorm(6); v
[1] -0.7984 -1.4916  0.6538 -0.3360 -1.6491  1.4902
> v < 0
[1]  TRUE  TRUE FALSE  TRUE  TRUE FALSE
```

And we can choose only those components of v that are non-negative:

```
> v[v >= 0]
[1] 0.6538 1.4902
```

The logical operators are

```
== equal to             != not equal to
< less than             <= less than or equal to
> greater than          >= greater than or equal to
```

1.11 Matrices

A matrix is a two-dimensional array of numbers, which may be constructed by a
recipe like the following:

```
> mat = matrix(c(80, 70, 55,   70, 50, 35),
nrow=2, ncol=3, byrow=TRUE)
> mat
      [,1] [,2] [,3]
[1,]   80   70   55
[2,]   70   50   35
```

nrow and ncol specify how many rows and columns there are, and byrow=TRUE
indicates that the numbers in c should be filled in by row. (The default is by column.) When the matrix is printed out, it is flanked by the row and column indices in
square brackets. Individual elements are specified by the row and column numbers
in square brackets:

```
> mat[1,2]
[1] 70
```

To get all the elements in row *nr*, write mat[nr,]. Likewise, to get all the
elements in column *nc*, write mat[,nc].

```
> mat[1,]
[1] 80 70 55
> mat[,2]
[1] 70 50
```

The row and column indices may be replaced by names to make a sort of table
structure. For example, suppose the rows in mat represent two drug treatments,
"Drug" and "Placebo", the columns represent the months after treatment, and the
numbers are the percent survivors in the two groups after the given treatment. Then
we could adapt the matrix structure as follows: [1]

```
> drug.test = matrix(c(80, 70, 55,   70, 50, 35),
  nrow=2, ncol=3, byrow=TRUE,
  dimnames = list(c("Drug","Placebo"),c("1","2","3")))
> drug.test
          1  2  3
Drug     80 70 55
Placebo  70 50 35
```

---

[1] Occasionally throughout this book it will be necessary to break a line of code or output, which
was originally written in a single line, into two or more lines to fit on the page. This is the case
with the > drug.test definition in the code block below. If you attempt to run such code, be
sure to remove the line breaks. In code that is originally broken into several lines, all lines but the
first are prefaced by a + sign.

1.11.1 Arithmetic operations on matrices

Define two matrices m1 and m2, each 2×2 for simplicity.

```
> m1 = matrix(c(1,2,3,4),nrow=2,byrow=T)
> m2 = matrix(c(1,1.1,1.2,1.3),nrow=2,byrow=T)
> m1
     [,1] [,2]
[1,]    1    2
[2,]    3    4
> m2
     [,1] [,2]
[1,]  1.0  1.1
[2,]  1.2  1.3
```

Just as with vectors, simple mathematical operations on the matrix are applied to each element individually.

```
> m1-3
     [,1] [,2]
[1,]   -2   -1
[2,]    0    1
> m1*5
     [,1] [,2]
[1,]    5   10
[2,]   15   20
> m1^2
     [,1] [,2]
[1,]    1    4
[2,]    9   16
> sin(m1)
        [,1]     [,2]
[1,] 0.8415   0.9093
[2,] 0.1411  -0.7568
```

If the two matrices are added (assuming their row and column dimensions are the same), the corresponding elements add.

```
> m1+m2
     [,1] [,2]
[1,]  2.0  3.1
[2,]  4.2  5.3
```

1.11.2 Matrix multiplication

If the two matrices are multiplied using the $*$ sign, each element in one is multiplied by the corresponding element in the other.

```
> m1*m2
      [,1]  [,2]
[1,]   1.0   2.2
[2,]   3.6   5.2
```

However, normal matrix multiplication is more complicated. The number of rows *nr* of the first matrix *m*1 must equal the number of columns *nc* of the second matrix *m*2. If that is the case, then the equation for the *ij*-th element of the product matrix *p* is

$$p_{ij} = \sum_{k=1}^{nr} m1_{ik} m2_{kj} \tag{1.1}$$

The symbol for matrix multiplication in R is $\%*\%$, just as for obtaining the dot product of two vectors. When this equation is applied to our two sample matrices,

```
> m1%*%m2
      [,1]  [,2]
[1,]   3.4   3.7
[2,]   7.8   8.5

> m2%*%m1
      [,1]  [,2]
[1,]   4.3   6.4
[2,]   5.1   7.6
```

Note that the result depends on the order of multiplication.

1.11.3 Determinant of a matrix

The determinant of a matrix is rarely needed in contemporary calculations, because computers enable more powerful and direct approaches. However, R has the det function if it is needed.

```
> det(m1)
[1] -2
```

1.11.4 Transpose of a matrix

The transpose of a matrix interchanges rows and columns. In R, the transpose function is denoted by t (matrix_name).

```
> t(m1)
     [,1] [,2]
[1,]    1    3
[2,]    2    4
```

1.11.5 Diagonal matrix

A diagonal matrix has non-zero elements on its diagonal. It can be constructed in R using diag.

```
> dmat = diag(1:3)
> dmat
     [,1] [,2] [,3]
[1,]    1    0    0
[2,]    0    2    0
[3,]    0    0    3
```

The $n \times n$ unit matrix, with 1s on the diagonal, can be constructed with diag(rep(1,n)).

```
> diag(rep(1,3))
     [,1] [,2] [,3]
[1,]    1    0    0
[2,]    0    1    0
[3,]    0    0    1
```

1.11.6 Matrix inverse

The inverse of a matrix M (it must be square) is denoted M^{-1}. Just as the inverse of 3 is 1/3, so that the product of a number and its inverse = 1, the matrix product of M and M^{-1} is the unit matrix I. The matrix inverse function in R is denoted solve because, as we shall see in Chapter 3 on Functions and Programming, the matrix inverse is often used to solve systems of linear equations.

```
> m1.inv = solve(m1)
> m1.inv
      [,1] [,2]
[1,] -2.0  1.0
```

```
[2,]   1.5 -0.5
> m1%*%m1.inv
           [,1] [,2]
[1,] 1.000e+00    0
[2,] 8.882e-16    1
```

Note that the [2,1] element is not exactly zero; it is zero only to within the numerical precision of the computer.

1.11.7 Eigenvalues and eigenvectors

If the product of a matrix A and a vector v yields v multiplied by a scalar λ,

$$Av = \lambda v \tag{1.2}$$

then we say that v is an eigenvector (or eigenfunction) of A and λ is the eigenvector corresponding to v. Looked at another way, if v is an eigenvector of A, then applying A to v simply multiplies the length of v by a constant λ, but does not change the direction of v (though the direction may be flipped if λ is negative). In R, the eigenvalues and eigenvectors of a matrix are obtained with the eigen function.

```
> eigen(m1)
$values
[1]   5.3723 -0.3723

$vectors
          [,1]     [,2]
[1,] -0.4160 -0.8246
[2,] -0.9094  0.5658
```

In general, an $n \times n$ matrix will have n eigenfunctions and eigenvalues. If you want just the eigenvalues without the eigenvectors, as is often the case, use the only.values = T option.

```
> eigen(m1,only.values=T)
$values
[1]   5.3723 -0.3723

$vectors
NULL
```

A more compact way to get this result is to use the formulation

```
> eigen(m1)$values
[1]   5.3723 -0.3723
```

Eigenvalues are used, for example, to determine whether the solutions to differential equations describing the competition of populations yield increasing, decreasing, or oscillating solutions. We shall see examples in subsequent chapters.

1.11.8 Other matrix functions

R has numerous other functions that operate on matrices, including `outer` (see Section 2.1.5) and `kronecker` (other forms of matrix multiplication), `svd` (singular value decomposition), `qr` (QR decomposition), and `chol` (Cholesky decomposition of positive symmetric matrices), You can learn about them by querying the R Help system with `help(function.name)` or `?function.name`, and by reading the "Arrays and Matrices" section of *An Introduction to R*.

There is also a `Matrix` package on the CRAN web site, which provides additional facilities for manipulating dense and sparse matrices. An introduction to this package is at
http://cran.r-project.org/doc/vignettes/Matrix/Introduction.pdf

Given this collection of matrix manipulation functions, R stands as a very capable platform for this important class of numerical analysis.

1.12 Other data structures

1.12.1 Data frames

In addition to the standard mathematical entities—numbers, vectors, and matrices—that we have discussed in this chapter, R has several other data structures that are key to working successfully with the program. The most important of these is *data frames*, which according to R Help are "tightly coupled collections of variables which share many of the properties of matrices and of lists, used as the fundamental data structure by most of R's modeling software." You can think of a data frame as a spreadsheet or flat database, with rows representing the individual cases and columns representing the fields (variables, factors, etc.) describing each case. We shall discuss data frames in some detail in Section 4.3.

1.12.2 Factors

Factors are numeric or character codes, arranged in a vector, that are used to categorize the individual members in a set of measurements. For example, a set of measurements of response of cells to a drug might be encoded with the factor "0" or "wt" if the cell is wild type, and "1" or "mut" if it is mutant.

1.12.3 Lists

The most general type of data structure in R is the *list*. To quote from R Help:

An R list is an object consisting of an ordered collection of objects known as its components.

There is no particular need for the components to be of the same mode or type, and, for example, a list could consist of a numeric vector, a logical value, a matrix, a complex vector, a character array, a function, and so on. Here is a simple example of how to make a list:

```
> Lst = list(name="Fred", wife="Mary", no.children=3,
            child.ages=c(4,7,9))
```

Components are always numbered and may always be referred to as such. Thus if `Lst` is the name of a list with four components, these may be individually referred to as `Lst[[1]]`, `Lst[[2]]`, `Lst[[3]]` and `Lst[[4]]`. If, further, `Lst[[4]]` is a vector subscripted array then `Lst[[4]][1]` is its first entry.

If `Lst` is a list, then the function `length(Lst)` gives the number of (top level) components it has.

Components of lists may also be named, and in this case the component may be referred to either by giving the component name as a character string in place of the number in double square brackets, or, more conveniently, by giving an expression of the form
`> name$component_name` for the same thing.

This is a very useful convention as it makes it easier to get the right component if you forget the number.

So in the simple example given above:

`Lst$name` is the same as `Lst[[1]]` and is the string "Fred",

`Lst$wife` is the same as `Lst[[2]]` and is the string "Mary",

`Lst$child.ages[1]` is the same as `Lst[[4]][1]` and is the number 4.

Sometimes it is not obvious whether a data structure is a numeric object such as a vector or matrix, a data frame, or a list. If the name of the structure is "Name", you can find out by typing `class(Name)`.

1.13 Problems

1. Use R as a calculator to find numerical answers to the following:

  a. $3+4(5+6)$
  b. 3^2+4^{1+2}
  c. $\sqrt{(1+2)(3+4)}$

    d. $\sqrt[3]{(3+4)(5+6)}$

    e. $\left(\frac{1.2+3}{4.5+6}\right)^2$

2. Use named variables to convert 2 years into months, days, minutes, seconds, and milliseconds.

3. Enter the numbers 3 5 7 9 11 13 into a data vector x. Using the vectorization property of mathematical operations,

    a. Find the square of each number.

    b. Find the square root of each number.

    c. Add 7 to each number.

    d. Subtract 6 from each number and then square the answers.

    e. Find the natural log and log to base 10 for each number.

    f. Find the sine of each number if its value is in radians.

    g. Find the sine of each number if its value is in degrees.

4. (From [65, p. 15]): You recorded your car's mileage at your last eight fill-ups as

    65311   65624   65908   66219   66499   66821   67145   67447

    Enter these numbers into the variable `gas`. Use the function `diff()` on the data. What does it give? Interpret what both of these commands return: `mean(gas)` and `mean(diff(gas))`.

5. (From [65, p. 15]): Four surveys between 1971 and 2000 showed the average number of calories consumed by a 20- to 29-year-old male to be 2450, 2439, 2866, and 2618. The percentage of calories from fat was 37%, 36.2%, 34%, and 32.1%. The percentage from carbohydrates was 42.2%, 43.1%, 48.1%, and 50%. Is the average number of fat calories going up or down? Is this consistent with the fact that over the same time period the prevalence of obesity in the United States increased from 14.5% to 30.9%?

6. Store the following data sets into a variable, using `c()` only when `:` or `seq()` will not work:

    a. The positive integers from 1 to 10

    b. The reciprocals of the positive integers from 1 to 10

    c. The squares of the integers from -5 to 5

    d. The years from 1990 to 2007

    e. The integers from 0 to 100 by 5

    f. 1, 2, 3, 5, 8, 13, 21, 34 (the first few terms of the Fibonacci series)

7. You make a series of ten measurements, obtaining the following results:

    20 19 23 27 25 18 24 18 20 25

    Use R functions to determine the minimum, maximum, and mean of these values. You then notice that the 27 was written down incorrectly; it should have been 21. Fix this, and find the new mean. Use R indexing and inequality operations (see Section 1.9.2, "Generating sequences of random numbers", to determine how many times the value was 23 or greater, and what fraction was less than 21.

8. Let y be the vector c(2,3,7,9,11,13). Guess the results of the R commands
   `length(y)`, `y>7`, `y[y>7]`, `mean(y>7)`, and `mean(y[y>7])`. Verify your
   guesses.

9. Construct a 3×3 matrix `mat` from the sequence 1:9, and a length-3 vector
   `vec = c(2,4,6)`. Perform the operations `vec*mat`, `mat*vec`, `mat%*%vec`,
   and `vec%*%mat`. Explain why your results are different or the same.

10. In a single R command, construct a 3×3 matrix `mat33` in which the 9 elements
    are drawn from the set of uniformly distributed random numbers between 0 and
    1.

Chapter 2
Plotting with R

An important part of scientific computing and data analysis is graphical visualiza-tion, an area in which R is very strong. R has many specialized graph types, some of which we will explore later. However, for many scientific purposes just a few types will suffice. We note especially graphs for data, functions, and histograms. We'll show simple examples of these first, and then show how they can be customized. For details on R graphics, see the book by Murrell [47].

2.1 Some common plots

2.1.1 Data plot

Graphing is particularly useful in descriptive statistics, to get an initial idea of how the data are distributed, whether their distribution is normal, etc. R makes it easy to do this.

We begin by generating a vector x of numbers from 0 to 100, and a vector $y = 3x - 2$, and then plotting y vs. x.

```
> x = seq(0,100, by=10)
> y = 3*x-2
> plot(x,y)
```

The simple `plot` command gives us a default graph with open circles as points, the names of the x and y variables as axis labels, and no title. To plot data with a line, instead of points, use `plot(x,y,type="l")`. See Section 2.2.2 for more variations on this theme.

A straight-line plot more relevant to biochemistry is the Lineweaver-Burk plot of Michaelis-Menten enzyme kinetics:

```
> Vmax = 10; Km=0.1 # Concentration in millimolar
> # Substrate concentrations
```

V. Bloomfield, *Computer Simulation and Data Analysis in Molecular Biology and Biophysics,*
Biological and Medical Physics, Biomedical Engineering, DOI: 10.1007/978-1-4419-0083-8_2,
© Springer Science + Business Media, LLC 2009

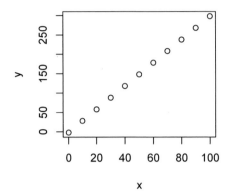

Fig. 2.1 Default data plot

```
> S = c(.01,.02,.05,.1,.2,.5,1,2,5)
> v = Vmax*S/(Km+S)  # Reaction velocity
> plot(1/S, 1/v)  # Lineweaver-Burk transformation
```

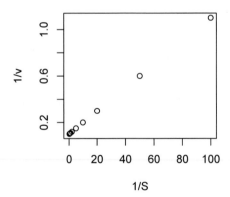

Fig. 2.2 Lineweaver-Burk data plot

2.1.2 Bar plot

Another common way of plotting data is a bar chart, which is known in R as a barplot. In the previous chapter we defined a matrix drug.test which gave the percentage survivors in two treatment groups after three time periods.

```
> drug.test = matrix(c(80, 70, 55,  70, 50, 35),
nrow=2, ncol=3, byrow=TRUE,
dimnames = list(c("Drug","Placebo"),c("1","2","3")))
> drug.test
          1  2  3
Drug     80 70 55
Placebo  70 50 35
```

We construct a barplot from this matrix with the following commands:

```
> barplot(drug.test,beside=TRUE,legend.text=TRUE,
  ylim=c(0,100),
main="Drug Test",xlab="Months",ylab="% Survival")
```

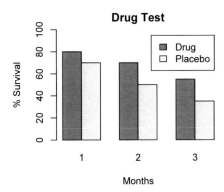

Fig. 2.3 Barplot of drug.test data

We have customized this plot by giving it a title with the `main` command, labeling the axes with `xlab` and `ylab`, and setting explicit limits on the y axis. We have also said that we wanted a plot in which the two treatment groups are plotted beside each other (the default, `beside = FALSE`), gives stacked bars; and that we wanted a legend (the default is `legend.text = NULL`). See ?barplot and Chapter 13 to see other options for `barplot`, as well as some more elaborate examples.

2.1.3 Function plot

To graph a function we use the `curve` function.

```
> curve(x*sin(x),-10,10, main="Function Plot")
```

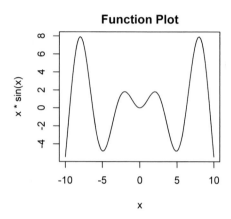

Fig. 2.4 Function plot of $x\sin(x)$ vs x

The expression to be graphed must be a function of x, or the name of the function to be plotted. The `-10,10` arguments are the `from`, `to` limits of the range. If they are in positions 2 and 3, just after `expr` in position 1, they don't need to be named explicitly. By default, `curve` evaluates the function at $n = 101$ points; that number can be changed as one of the options of `curve()`.

2.1.4 Histogram

Histograms are important to visually inspect the distribution of repeated measurements. For example, we generate 100 normally-distributed random numbers with mean = 2 and standard deviation = 0.2, and plot them using the `hist` function

```
> y = rnorm(100,mean=2, sd=.2)
> hist(y)
```

The title "Histogram of y" is automatically added by the `hist` function. You can specify a different title with `main = "Desired Title"`.

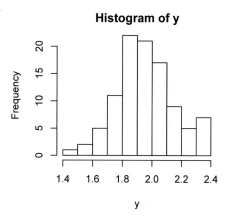

Fig. 2.5 Histogram of 100 normally-distributed random numbers

2.1.5 Three-dimensional plot

Sometimes we want to plot a dependent variable as a function of two independent variables. The R function persp gives such a plot with a customizable perspective that enables us to get the most informative view. A good example (simplified a bit to print in black and white) is given in R Help for persp. We first view z from the front as a function of x and y, and then rotate it by 30 degrees in the horizontal and vertical directions. (User-defined functions will be discussed in the next chapter.)

```
x = seq(-10, 10, length= 30)
y = x
f = function(x,y) { r = sqrt(x^2+y^2); 10 * sin(r)/r }
z = outer(x, y, f) # Forms matrixz using  function f
par(mfrow = c(1,2))
persp(x,y,z)
persp(x,y,z,theta = 30, phi = 30)
```

Consult R Help to learn the many options available to customize the appearance of the plot produced by persp.

The value of z as a function of x and y can also be visualized by a contour map or a two-dimensional image colored according to the value of z, as in a topographical map. See contour and image in R Help for information on these functions.

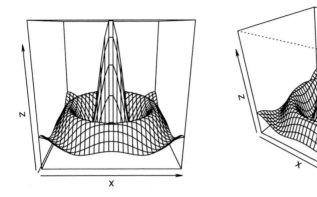

Fig. 2.6 Perspective plots. (left) Front view; (right) rotated by 30 degrees about θ and ϕ angles

2.2 Customizing plots

2.2.1 Different plot characters

By default, R plots points with open circles, but many other symbols are available. For example, to plot with closed circles, use pch=16, where pch stands for "plot character". Here's a Lineweaver-Burk plot with closed circles.

```
> plot(1/S, 1/v, pch=16)
```

Other point types are available using pch values between 0 and 25. The default (open circles) is pch = 1. Here is some code (don't worry if you don't understand it now) that shows the various point characters (pch) and line types (lty) available.

```
> plot.new()
> plot.window(xlim=c(-.5,26.5),ylim=c(0,8), asp=1)
> k = 0:25
> zero = 0*k
> text(k,8+zero, labels=k)
> points(k,7+zero,pch=k,cex=2)
> i = 6:1
> abline(h=7-i,lty=i)
> axis(2,at=1:8,labels=c(paste("lty =",i),"pch","k"), las=2)
```

In most circumstances you would use the simplest plot characters, 1–3 and 15–17.

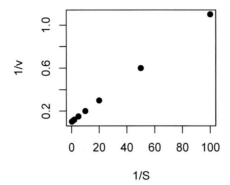

Fig. 2.7 Lineweaver-Burk plot with filled circles as data points

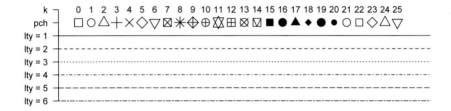

Fig. 2.8 Plot symbols and line types

The size of the plot characters can be controlled with cex. The default size is cex=1. Point sizes can be halved with cex=0.5, or made 50% larger with cex=1.5, etc.

```
> x = seq(0,100, by=10)
> y = 3*x-2
> plot(x,y, pch=17, cex=2)
```

2.2.2 Plotting data with a line

To plot with a line, use type = "l". (To plot with points, you can also explicitly specify type = "p".) To get both points and a line, type = "o" (for over-strike), or type = "b" for both.

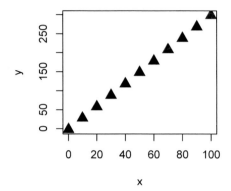

Fig. 2.9 Enlarged plot characters

```
> par(mfrow=c(1,2))
> plot(x,y,pch=16, type="o")
> plot(x,y,pch=16, type="b")
```

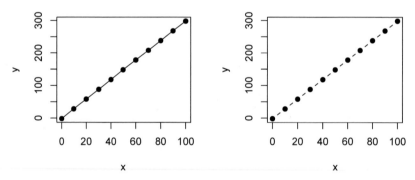

Fig. 2.10 Combining points and lines. (left) "o"; (right) "b"

In either data plots or function curves, you can specify the line type with lty = n, where *n* ranges from 1 to 6 (see above). You specify the line width with lwd. For example, here's a plot of the initial velocity of an enzyme according to Michaelis-Menten kinetics. . You can thicken the line , or plot with dots rather than a solid line.

```
> Vmax = 10; Km = 0.1 # Concentration in millimolar
> S = c(.01,.02,.05,.1,.2,.5,1,2,5)
> v = Vmax*S/(Km+S)
```

```
> plot(S,v,type="l")
> plot(S,v,type="l",lwd=2)  # Thicker line
> plot(S,v,type="l",lty=3)     # Dotted line
```

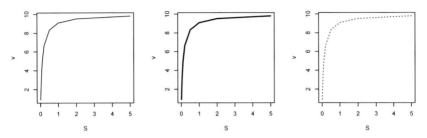

Fig. 2.11 Plotting data points with a (left) line, (center) thicker line, (right) dotted line

2.2.3 Adding title and axis labels

To put a title on the graph, use "main". (If a subtitle below the graph is desired, add it with "sub".) To change the x- and y-axis labels from the defaults, use "xlab" and "ylab". (We already did this for the barplot example above.) Be sure to surround the names with quotation marks.

```
> plot(S,v,main="Michaelis-Menten Kinetics",
  xlab="S/mM", ylab="v/(mM/s)")
```

If a graph can be produced without title, subtitle, legend, or axis labels, these items can be added later (while the graph is still active) with the title command. See ?title in R Help for details.

2.2.4 Adding colors

To change the colors of points or lines, use the col command:

```
> x = 1:10
> y = x^3-2*x^2+4*x-1
> plot(x,y,pch=16,cex=1.3,col="red")
```

Since this book is printed in black and white, we do not show the resulting colored graph.

You can specify colors either by name (in quotes) or by number. The first four are (1) "black", (2) "red", (3) "green", and (4) "blue". You can learn more

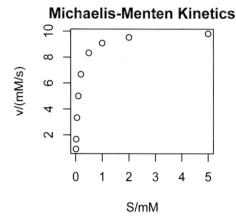

Fig. 2.12 Placing a title above the graph

about the colors available in R, and how to use them, by typing ?colors. Note that in a typical graph for publication, you would generally stick to black and white, distinguishing data series by point or line type. However, for a poster or computer-based presentation, colors are helpful.

To customize colors or shading of barplots, as well as many other barplot features, see ?barplot in the R help system.

2.2.5 Adding straight lines to a plot

You can add straight lines to a plot with the abline function. This can take several forms. If you want to draw a line with specified intercept a and slope b (hence the name), use abline(a,b). For example, to draw a horizontal dashed line (slope b = 0) on the function plot at the beginning of this chapter, to estimate where the function crosses the y axis (intercept a = 0):

```
> curve(x*sin(x),-10,10, main="Function Plot")
> abline(0,0,lty=2)
```

Note that you add the line to the already drawn basic curve. This method of sequentially adding details to a graph, rather than specifying everything at once initially, is standard in R. But the graph must be "open", the latest one drawn.

Another variation on abline is to use it to draw a vertical or horizontal line, using "v" and "h", respectively. For example, to draw a vertical line through the value of S corresponding to the Michaelis constant K_m, and a horizontal line through the value of v corresponding to the half-maximum velocity $V_{max}/2$:

```
> # Kinetic parameters
> Vmax = 10; Km = .001
```

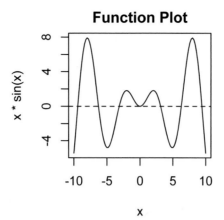

Fig. 2.13 Adding a line to the graph with abline

```
> # Substrate concentrations
> S = c(.0001,.0003,.001,.002,.005,.01,.02)
> v = Vmax*S/(Km+S)   # Calculated velocities
> plot(S,v,type="b")
> abline(v=Km,lty=3) # Vertical line at Km
# Don't confuse v here, which means "vertical",
# with the enzyme velocity
> abline(h=Vmax/2, lty=3)  # Horizontal line at Vmax/2
```

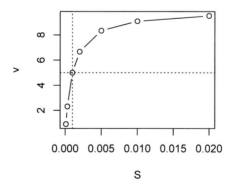

Fig. 2.14 Adding horizontal and vertical lines with abline to locate the half-maximum velocity and Michaelis constant

2.2.6 Adjusting the axes

R automatically sets the upper and lower limits of the horizontal and vertical axes. Sometimes you may want to use different limits, which you can do with xlim=c(xlo,xhi) and ylim=c(ylo,yhi). For example, in the Lineweaver-Burk plot of the simulated data above,

```
> plot(1/S, 1/v, pch=16, main="Lineweaver-Burk Plot",
  ylim=c(0,1.2))
# Draw line corresponding to intercept and slope of
# Lineweaver-Burk plot
> abline(1/Vmax,Km/Vmax, lty=3)
```

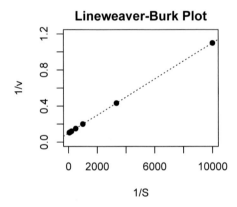

Fig. 2.15 Lineweaver-Burk plot with axes adjusted and line added to show slope and intercept

2.2.7 Customizing ticks and axes

By default, R draws the axis ticks outside the graph. Most commonly in scientific journals, the ticks are inside. This can be set by a command like tck=0.03, where the positive value of tck means ticks inside the graph, and 0.03 is the length of the ticks relative to the graph.

```
> plot(1/S, 1/v, pch=16, tck=0.03,
  main="Lineweaver-Burk Plot",ylim=c(0,1.2))
> abline(1/Vmax,Km/Vmax, lty=3)
```

R's choice of ticks on the axes is usually adequate. To learn how to customize, call ?axis in the R help system. The axis documentation will also tell you, among other things, how to draw additional axes on the top- or right-hand side of

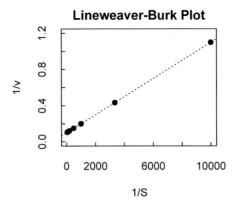

Fig. 2.16 Graph with customized ticks

the plot, using the `side` option. This would be useful, for example, if you wanted to plot two variables which had significantly different magnitudes. The right-hand axis (4) could show the scale for the second variable.

2.2.8 Setting default graph parameters

Rather than setting graph parameters individually for each graph, if you have a preferred style you can set it with the `par` command. For example, to generate default plots with ticks inside and filled circles, use

```
> par(tck=0.03, pch=16)
```

Every graph that you draw from then on until you quit the session will use these defaults, unless you change or override them. Of course, settings for a particular graph can be changed as above. `par` has many options, as you can see from reading R Help, that enable you to customize many aspects of a plot layout.

2.2.9 Adding text to a plot

You may wish to annotate a plot by adding text within it. For example, you can place labels indicating the intercept $1/V_{max}$ and slope K_m/V_{max} on the Lineweaver-Burk plot from the data at the beginning of this chapter as follows:

```
> text(15,0.5,"1/Vmax")
> text(60,0.5,"Km/Vmax")
```

To add text in the margins of a plot, rather than within the plotting area, use `mtext`; see R Help for details. If there are several data series or symbols in a graph,

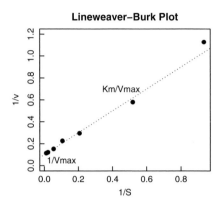

Fig. 2.17 Graph with added text and customized graph parameters

one can annotate with a legend. We'll see examples of this below. Plots can also be annotated with mathematical symbols and expressions. To learn about this, call `?plotmath` in R.

2.2.10 Adding math expressions and arrows

R has useful capabilities for adding mathematical symbols and expressions to a plot, and for pointing out significant features with arrows. As a simple example, we plot the function $\sin(x)/x$ which appears in x-ray diffraction, give the equation within the plot, and draw an arrow to indicate a significant maximum.

```
> x = seq(-20,20,by=0.1)
> y = sin(x)/x
> plot(x,y, type = "l")
> text(-15,0.7,expression(frac(sin(x),x)))
> arrows(11,0.3,8,0.15, length=0.1)
> text(11,0.45,"2nd\nmaximum") # \n means "newline"
```

For details, see `plotmath`, `demo(plotmath)`, and `arrows`. As we'll see in the section on error bars at the end of this chapter, arrows are also used to draw error bars.

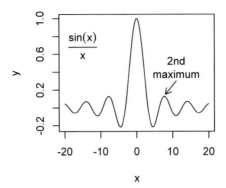

Fig. 2.18 Plot with added math expressions and arrows

2.2.11 Constructing a diagram

Sometimes you will want to construct a block diagram, organization chart, or similar figure composed mainly of straight lines, arrows, and text. R, with its large number of graphic constructs, enables you to do this fairly readily. As an example, we consider the code used to construct the diagram for countercurrent distribution in Chapter 8. Comparing the commented code with the diagram should illuminate most of the steps, but several points are worth noting.

- It is useful to define variables that specify the dimensions, so that sizes and proportions can be changed in one place rather than in each line of code.
- The `plot.new()` function defines a new plot without axes, labels, or outlining box. The `plot.window()` function sets the horizontal and vertical scales.
- The `segments(x0,y0,x1,y1)` function draws straight lines from `x0,y0` to `x1,y1`. It uses the same syntax as the `arrows` function, but does not include arrowhead length and angle. Note the use of `for...` loops to place repetitive vertical lines.
- Subscripts in plotted text are indicated by square brackets. See `plotmath`, `expression`, and `substitute` in R Help for details and other possibilities for math notation. The `tH` and `tL` functions are defined for brevity and convenience in writing the commands to place text on the high and low levels.

```
# Assume cells with square sides of length 1
# Assign variables to the dimensions in the diagram.
lmarg = 1; rmarg = 1 # Margins for text and arrows
n1 = 4; n2 = 1.5; n3 = 2 # Lengths of horizontal walls
nv = 0.3 # Lengths of vertical walls
pw = lmarg + n1 + n2 + n3 + rmarg # Plotwidth
ph = 2 # Plotheight
```

```
# Prepare a new, empty plot and dimension its window
plot.new()
plot.window(c(0,pw),c(0,ph))

# Draw straight lines using segments
# Lower horizontals
# Lower left
segments(lmarg,0,lmarg + n1,0)
# Lower middle
segments(lmarg + n1,0,lmarg + n1 + n2,0,lty=2)
# Lower right
segments(lmarg + n1 + n2,0,lmarg + n1 + n2 + n3,0)
# Upper horizontals
# Upper left
segments(lmarg,2,lmarg + n1,2)
# Upper middle
segments(lmarg + n1,2,lmarg + n1 + n2,2,lty=2)
# Upper right
segments(lmarg + n1 + n2,2,lmarg + n1 + n2 + n3,2)
# Middle horizontals
# Middle leftsegments(lmarg,1,lmarg + n1,1, lty=3)
# Middle middle
segments(lmarg + n1,1,lmarg + n1 + n2,1,lty=2)
# Middle right
segments(lmarg + n1 + n2,1,lmarg + n1 + n2 + n3,1,
  lty=3)
# Left verticals
for (i in 1:(n1+1)){
segments(lmarg + i -1,2,lmarg + i -1,2-nv)
segments(lmarg + i -1,1+nv,lmarg + i -1,1-nv)
segments(lmarg + i -1,nv,lmarg + i -1,0)
}
# Right verticals
for (i in 1:(n3+1)){
segments(lmarg + n1 + n2 + i -1,2,
  lmarg + n1 + n2 + i -1,2-nv)
segments(lmarg + n1 + n2 + i -1,1+nv,
  lmarg + n1 + n2 + i -1,1-nv)
segments(lmarg + n1 + n2 + i -1,nv,
  lmarg + n1 + n2 + i -1,0)
}

# Annotate diagram with text, including subscripts
# For brevity, define functions for high and low text
tH = function(x,i) text(x, 1.5, substitute(H[i]))
```

```
tL = function(x,i) text(x, 0.5, substitute(L[i]))
# Upper
tH(lmarg-0.8,"in"); tH(pw-rmarg+0.8,out)
tH(1.5,1); tH(2.5,2); tH(3.5,3); tH(4.5,4)
tH(n1+n2+1.5,n-1); tH(n1+n2+2.5,n)
# Lower
tL(lmarg-0.8,out); tL(pw-rmarg+0.8,"in")
tL(1.5,1); tL(2.5,2); tL(3.5,3); tL(4.5,4)
tL(n1+n2+1.5,n-1); tL(n1+n2+2.5,n)

# Draw arrows for in- and out-flows
arrows(lmarg-0.4,1.5,lmarg,1.5,.1,30)
arrows(pw-rmarg,1.5,pw-rmarg+0.4,1.5,.1,30)
arrows(lmarg,0.5,lmarg-0.4,0.5,.1,30)
arrows(pw-rmarg+0.4,0.5,pw-rmarg,0.5,.1,30)
```

2.3 Superimposing data series in a plot

Often you'll want to plot two sets of data on the same graph, or two functions, or
data points and a function. To do this, graph one of the series first, then the other.
You can then add a legend to identify the series. It's important to note that, for this to
work, the axis limits of the first-drawn graph must be large enough to accommodate
the subsequent series. For example,

```
> x = 1:10
> y1 = x^2-3*x+2
> y2 = x^2-2*x+3
> plot(x,y1, pch=16)
# Use points(), not plot(), to add the second series
> points(x,y2,pch=1)
```

The problem with this graph (Figure 2.19) is that the highest value of y2 is
greater than the axis limit set by y1, so the last point is cut off. To avoid this, first
test the minima and maxima of the series using range(), then set the axis limits
accordingly. Note also that the y axis is labeled "y1", which probably should be
changed to the generic "y".

```
> range(y1)
[1]  0 72
> range(y2)
[1]  2 83
> plot(x,y1,ylim=c(0,90), ylab="y",pch=16)
> points(x,y2,pch=1)
```

A more general way to set ylim would be to use min(c(y1,y2)) for the
minimum value (or perhaps min(c(0,y1,y2)) if you wanted the plot to start at 0

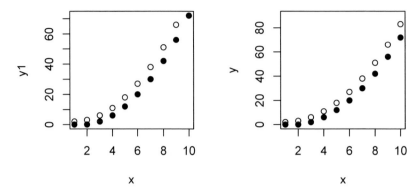

Fig. 2.19 Plotting two data sets on the same graph. (left) original, (right) after adjustment of *y* axis limits

and knew that no y values were below 0), and `max(c(y1,y2))` for the maximum value. Combining,

`ylim = c(min(c(0,y1,y2)),max(c(y1,y2))).`

The function `matplot` provides a more concise way to superimpose data series. `matplot` plots the columns of a matrix individually as a function of x. The matrix can be assembled using `cbind`. Using the example above, `cbind(y1,y2)` is a 10×2 matrix, and x is a length-10 column vector. If you simply enter the command `matplot(x,cbind(y1,y2))` you will get a graph in which the points in the two data series are denoted by the numbers 1 or 2, and in which 1 is printed in black (`col=1`) and 2 in red (`col=2`). (Try it.) To get a plot like the one above, all in black and with point characters, you must use something like

```
> matplot(x,cbind(y1,y2),type="p",pch=c(16,1),
    col=c(1,1), ylab="y")
```

or

```
> y = cbind(y1,y2)
> matplot(x,y,type="p",pch=c(16,1),col=c(1,1))
```

Note that `matplot` automatically sets `ylim` to include both data series.

To identify the series, add a legend with the `legend` function, which specifies the *x* and *y* coordinates of the legend, its text, accompanying symbols or line types, and whether the legend is surrounded by a box. To suppress the box, use `bty="n"` where `bty` stands for "box type".

```
> legend(1,85,legend=c("y1","y2"),pch=c(16,1),bty="n")
```

To plot a line (for example, a fitting function) on a data plot, first plot the points, then the line with the function `lines`, then the legends. You may have to experiment a bit to get the legends located properly.

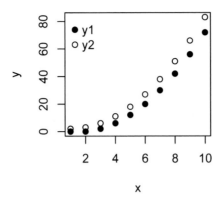

Fig. 2.20 Matplot with legend

```
> x = 1:10
> y1 = x^2-3*x+1 +rnorm(10,0,5)
> plot(x,y1)
> lines(x^2-3*x+1)
> legend(2,60,legend="y1",pch=16,bty="n")
> legend(1.5,50,legend="fit",lty=1,bty="n")
```

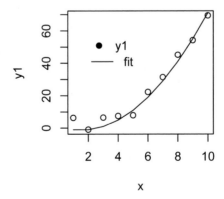

Fig. 2.21 Data plot with superimposed fitting curve

To plot two or more curves, use the add = TRUE option to the second and later curve command. For example,

```
> curve(x^3-3*x^2+2*x-4,-10,10,
```

```
   ylim=c(-1500,1500),ylab="f(x)")
> curve(x^3+3*x^2+2*x-100,-10,10,
   lty=2, add=TRUE)
> legend(-9,1000,legend=c("f1(x)","f2(x)"),bty="n",
+ lty=c(1,2))
# "+" is line continuation in R
```

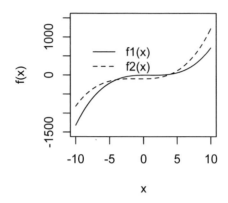

Fig. 2.22 Two superimposed curves with legend

2.4 Placing two or more plots in a figure

R provides several methods for placing several plots in a single figure. The most common method is par(mfrow=c(nr,nc)) where nr is the number of rows of plots and nc is the number of columns. mf stands for "multiple figures". With mfrow, the plots are successively filled in row order. The alternative mfcol fills the plots in column order.

For example, consider Michaelis-Menten enzyme kinetics with values of the reaction velocity v calculated from the substrate concentration S and the kinetic parameters V_{max} and K_m. We plot v vs. S and the three common linear transformations of the data in a single figure as follows.

```
> S = c(0.1,0.2,0.5,1,2,5,10,20)*1e-6
> Km = 2e-6
> Vmax = 10
> v = Vmax*S/(Km+S)
> par(mfrow=c(2,2)) # Set up 2 x 2 plot figure
> plot(S,v,type="o", main="Michaelis-Menten")
```

```
> plot(1/S,1/v, type="o", main="Lineweaver-Burk")
> plot(v/S,v,type="o", main="Eadie-Hofstee")
> plot(S,S/v, type="o", main="Hanes-Woolf")
> par(mfrow=c(1,1)) # Return to single plot
```

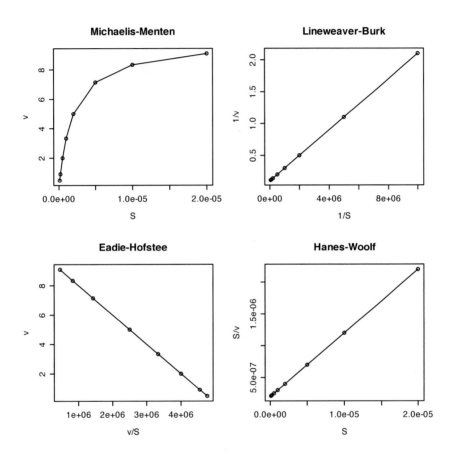

Fig. 2.23 Four plots in the same figure

Note that at the end we used par(mfrow=c(1,1)) to return the graphic parameters to a single plot format. This is necessary since par commands stay in effect until changed.

There is a good deal of space between the plots in this figure. It might be desired to compress this space, especially if the individual plots did not have titles. This can be done with the mar option to par. As explained in R Help for par, mar is "A numerical vector of the form c(bottom, left, top, right) which gives the number of lines of margin to be specified on the four sides of the plot. The default is c(5, 4, 4, 2) + 0.1." To compress the space between the plots, these values could be decreased.

mfrow and mfcol divide the figure into equal-sized regions; all plots are the same size. If it is desired to have different-sized plots, R provides the functions layout and split.screen. Consult R Help to see how these work.

2.5 Error bars

We use error bars in graphs to denote statistical variability or uncertainty. To add error bars, we use the arrows function in R. According to the documentation for that function in the help system,

```
Sample usage:

arrows(x0, y0, x1, y1, length = 0.25, angle = 30)

Arguments:

x0, y0: coordinates of points from which to draw.
x1, y1: coordinates of points to which to draw.
length: length of the edges of the arrow head (inches).
angle: angle from the shaft of the arrow to the edge
    of the arrow head.
```

There are several other arguments that we won't need. Here's an example of a plot of ten uniformly distributed numbers, with error bars equal to $\pm 10\,\%$ of the y value:

```
> x = 1:10
> y = runif(10)
> plot(x,y)
> arrows(x,y,x,y+.1*y,.05,90); arrows(x,y,x,y-.1*y,
    .05,90)
```

Here's an example of a bar chart with error bars, with data from six uniformly distributed random numbers between 0 and 1. This example also shows how to generate cross-hatching of the bars at a density of 15 lines per inch and a black color (medium gray is the default), to set the counterclockwise angle of the hatching to 45 degrees, and to show error bars only above the top of the bar, as is the common convention.

```
> x = runif(6)
> bar = barplot(x, names.arg = month.abb[1:6],
+ density = 15, angle = 45, col="black", ylim=c(0,1.1))
> stdev = x/10
> arrows(bar, x, bar, x + stdev, length=0.1,
    angle = 90)
```

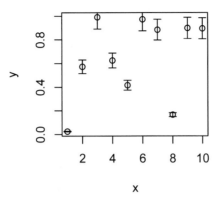

Fig. 2.24 Data plot with error bars

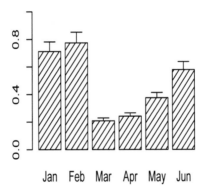

Fig. 2.25 Bar chart with cross-hatching and error bars

The labels are generated from the built-in R values for the three-letter abbreviations for the months (`month.abb`). The only other R built-in constants are the full names of the months (`month.name`), the upper-case (`LETTERS`) and lower-case (`letters`) letters, and `pi`.

2.6 Locating and identifying points on a plot

R has functions that enable identifying and placing points on a plot, and might be useful for digitizing a plot. The function `locator`, which "reads the position of the graphics cursor when the (first) mouse button is pressed" can usefully serve as a digitizer for a plot. For example, the command `xy = locator(10)` followed by ten mouse clicks on an experimental curve or set of points, will yield a set of 10 (x, y) coordinates, which can be accessed by `xy$x` and `xy$y`. If desired, `locator` will put a point at each point clicked, or draw a line between clicked points.

A related function is `identify`. According to the R documentation, "`identify` reads the position of the graphics pointer when the (first) mouse button is pressed. It then searches the coordinates given in x and y for the point closest to the pointer. If this point is close enough to the pointer, its index will be returned as part of the value of the call."

See R Help for details on how to apply these functions.

2.7 Problems

1. The BOD data in the R "datasets" package gives biological oxygen demand (mg/l) vs time (days) in an evaluation of water quality.

   ```
   time = c(1, 2, 3, 4, 5, 7)
   demand = c(8.3, 10.3, 19.0, 16.0, 15.6, 19.8)
   ```

   Plot demand vs time, using filled black circles as the points. Give the plot an appropriate title.
2. Repeat Problem 1, with the following modifications: Change the minimum value of the y-axis to 0. Change the x-axis label to "days" and the y-axis label to "BOD". Move the ticks to the inside of the graph. Connect the black points with a solid red line.
3. Assume that the demand measurements have $\pm10\%$ error. Repeat Problem 2, with error bars on the points.
4. Assume that the water was treated to reduce BOD, and a second set of measurements gave the following results:

   ```
   time = c(1, 2, 3, 4, 5, 6, 7, 8)
   treated = c(5.1,7.7,10.2,12.3,14.0,13.3,15.5,13.9)
   ```

   Assume $\pm10\%$ error for these data as well. Plot both untreated and treated data sets on the same graph, using different point and line types and/or colors for each plot. Add a legend (unboxed) that labels the data.
5. Add horizontal lines to the plot in Problem 4 to represent the mean values of the BODs in the two measurements. Make the line types the same as those that connect the points.
6. You guess that an approximate functional approximation to the data in Problem 4 is

   ```
   treated = 5 + 8*time/(2 + time)
   ```

   Add this line to a plot of the treated data.

7. Plot the untreated and treated BOD data as a barplot, with appropriate legend and title. Since data for 6 and 8 days are missing from the untreated set, use NA for those values. Make the untreated data solid blue, and the treated data solid red. See ?barplot in R Help.

8. Modify the plot in Problem 7 to make the untreated bars shaded gray at 45 degrees counterclockwise, and the treated bars 45 degrees clockwise. Add error bars to both sets of bars.

9. Generate a vector rn of 1000 normally distributed random numbers with mean 0 and standard deviation 1. Use the command par(mfcol=c(1,2)) to tell R to plot two graphs side-by-side. (See R Help on ?par.) In the left-hand graph plot the histogram of rn; in the right-hand graph plot hist(rn, freq=F). Note the difference in ordinate.

10. While still placing two graphs side-by-side, compare hist(rn, breaks=10) with hist(rn, breaks=20). Put a title on each graph so you can tell them apart. Do you see a difference in the number of breaks? Consult R Help about ?hist to learn how to control the number of breaks. Repeat the two graphs using that approach. Remember to reset the graphing system to one row, one column when you are done.

Chapter 3
Functions and Programming

3.1 Built-in functions in R

The base installation of R has many built-in functions, including `sort` and `order` to arrange vectors or arrays in ascending or descending order; all the standard trigonometric and hyperbolic functions `log`(base e), `log10`, `exp`, `sqrt`, `abs`, etc.; and more sophisticated mathematical functions such as `factorial`, `gamma`, `bessel`, `fft` (Fourier transform), etc. Additional mathematical functions, the orthogonal polynomials used in mathematical physics and chemistry, are available in the contributed package `orthopolynom`, available through the CRAN web site. The functions `uniroot` (Section 3.4.1) and `polyroot` (Section 3.4.2) are used to solve for the zeros of general functions and polynomials, respectively.

In addition to those mathematical functions, R has numerous others that are useful to scientists. We'll discuss three of them here in this section: sorting, splines, and sampling. At the end of this chapter, we'll introduce some of the R functions that are commonly used in numerical analysis.

A few general observations about built-in functions in R: Their arguments often have defaults, which need not be specified if they're satisfactory, while other arguments do need to be specified. For example, in `plot(x,y)` the arguments x and y must be specified, while the point type (open circle) and x and y limits are defaults which can be changed.

Also, the arguments in a function may be called either by name or by position. For example, the function for generating regular sequences is formally called by

```
seq(from = 1, to = 1, by = ((to - from)/(length.out
  - 1)), length.out = NULL, along.with = NULL, ...)
```

However, we can also call it simply by the position of the arguments, e.g., `seq(1, 10,0.5)`. Note also that if we want to specify the `length` of the sequence, rather than the by step, we can type `seq(1,10,length = 19)` or even `seq(1,10, l=19)` rather than `seq(1,10, length.out = 19)`. The argument name need not be completely spelled out so long as there is no ambiguity with the names of other arguments.

V. Bloomfield, *Computer Simulation and Data Analysis in Molecular Biology and Biophysics,*   49
Biological and Medical Physics, Biomedical Engineering, DOI: 10.1007/978-1-4419-0083-8_3,
© Springer Science + Business Media, LLC 2009

3.1.1 Sorting

If we want to sort just a single vector, we use `sort`. For example, we sort a vector
of six uniformly distributed random numbers.

```
> rnum = runif(6)
> rnum
[1] 0.70133 0.95279 0.48310 0.04124 0.78296 0.45833
> sort(rnum)
[1] 0.04124 0.45833 0.48310 0.70133 0.78296 0.95279
> sort(rnum, de = TRUE)
[1] 0.95279 0.78296 0.70133 0.48310 0.45833 0.04124
```

The default is to sort in ascending order. To sort in descending order, we use the
option `descending = TRUE`. Since there is no ambiguity with the name of another option, "descending" can be abbreviated. If there are NA values in the vector,
the default is to sort them to the end (`na.last = TRUE`). If it is desired to put
NA values first, use `na.last = FALSE`.

To sort a data frame, in which all the columns need to be rearranged in step with
the one being sorted, use `order`. For example, we define a character vector `lets`
and a logical vector `tf` and combine them with the numerical vector `rnum` to form
a data frame.

```
> lets = c("e", "f", "o", "u", "c", "r")
> tf = c(T,F,T,T,F,T)
> datf = data.frame(rnum, lets, tf)
> datf
     rnum lets      tf
1 0.70133    e   TRUE
2 0.95279    f  FALSE
3 0.48310    o   TRUE
4 0.04124    u   TRUE
5 0.78296    c  FALSE
6 0.45833    r   TRUE
```

Then if we order on `rnum`, using the notation `datf$rnum` to pick the desired
vector, we get

```
> datf1 = datf[order(datf$rnum), ]
> datf1
     rnum lets      tf
4 0.04124    u   TRUE
6 0.45833    r   TRUE
3 0.48310    o   TRUE
1 0.70133    e   TRUE
5 0.78296    c  FALSE
2 0.95279    f  FALSE
```

If, instead, we order on `lets` in decreasing order, we get

```
> datf2 = datf[order(datf$lets, decreasing = TRUE), ]
> datf2
      rnum lets      tf
4 0.04124    u   TRUE
6 0.45833    r   TRUE
3 0.48310    o   TRUE
2 0.95279    f  FALSE
1 0.70133    e   TRUE
5 0.78296    c  FALSE
```

3.1.2 Splines

Splines are used to fit a smoothing curve through a set of points. For example,

```
> x = 1:10
> y = rnorm(10)
> plot(x,y)
> lines(spline(x,y))
```

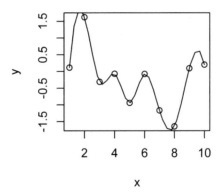

Fig. 3.1 Smoothing spline curve through a set of points

The default is for `spline` to draw the curve through three times as many inter-polating points as there are original points. If there are relatively few original points, the curve can appear jerky, so you can set a larger number of interpolating points.

`spline(x,y)` gives the actual x and y values of the interpolating points. These, or more conveniently the interpolating function `splinefun(x,y)`, can be used for calculations such as numerical integration and differentiation, as we'll see below (Sections 3.4.6, 3.4.7). See `?spline` for details and examples.

Related functions are `approx` and `approxfun`. The first returns a list of points which linearly interpolate given data points, while the second gives a function that performs the linear (or constant) interpolation. Consult `?approx` for more details.

3.1.3 Sampling

A common operation in computer simulation is to sample a distribution by picking members of the distribution at random. Familiar examples are tossing a coin and picking a DNA base if the fractions of the four bases are known. The R function `sample` takes a sample of the specified size from the elements of a vector x, either with or without replacement, using specified probabilities. For example, to pick ten DNA bases at random if the probability of choosing each of the bases is 0.25, use

```
sample(c("A","C","G","T"), size = 10, replace = TRUE,
    prob = c(0.25, 0.25, 0.25, 0.25))
```

Note that you must specify `replace = TRUE`, otherwise once a base has been chosen, it will no longer be available.

3.2 User-defined functions

However, there will be times when you'll want to define your own functions. In general, the syntax for defining a function f of variables x, y, ... is
`f = function(x,y,...) {expression involving x, y, ...}`
Functions can yield either the results of a relatively simple mathematical expression, such as the Gaussian function below, or a multistep procedure (what some other programming languages call a subroutine). In either case, the value returned by the function is the value of the last operation in the expression defining the function. For example, if the function separately calculates x, y, and z, and you want $x+y+z$, the last operation in the function should be `return(x+y+z)` See *10. Writing your own functions* in R Help for more information. Several specific examples are given below.

3.2.1 Gaussian function

As a first example, consider the familiar bell-shaped curve, or Gaussian, or normal distribution. We'll call the function `gauss`, and define it as follows:

```
> gauss = function(x0,x,sig) {1/(sqrt(2*pi)*sig)*
    exp(-(x-x0)^2/(2*sig^2))}
```

where x is the variable, $x0$ is the maximum, and *sig* is the standard deviation. The factor of `1/(sqrt(2*pi)*sig)` is the normalization constant, so that the total area under the curve = 1. Then we can plot an instance of the function, centered at $x0 = 10$ and with $sig = 2$, simply by writing

```
curve(gauss(10,x,2),0,20, main="Gaussian Function")
```

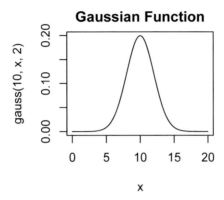

Fig. 3.2 Curve plot of a Gaussian function

It is now easy to plot a sum of Gaussians, as might occur in the UV spectroscopy of a protein or nucleic acid.

```
> curve(gauss(280,x,20)+.5*gauss(210,x,20)+
  .2*gauss(190,x,10),150,350,xlab="nm", ylab="OD",
  main="Sum of 3 Overlapping Spectra")
```

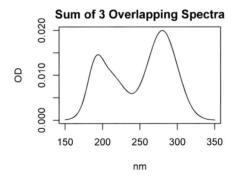

Fig. 3.3 Sum of three overlapping Gaussian curves

We can plot the three spectral components separately on the same graph, in black, red, and blue, and then plot the sum in green.

```
> curve(gauss(280,x,20),150,350,xlab="nm",ylab="OD",
  main="Decomposition of Spectrum")
> curve(.5*gauss(210,x,20), add=TRUE, col="red")
> curve(.2*gauss(190,x,10), add=TRUE, col="blue")
> curve(gauss(280,x,20)+.5*gauss(210,x,20)+
  .2*gauss(190,x,10), add=TRUE,col="green")
```

3.2.2 pH titration curves

As a second example, we'll consider pH titration curves and the charge on a peptide. Begin with the Henderson-Hasselbalch equation

$$pH = pK_a + \log_{10} \frac{f_B}{1 - f_B} \tag{3.1}$$

We define two functions, fB to calculate the fraction of a moiety with a given pK_a in the basic form at a given pH, and qNet to calculate the net charge on that moiety given the charge q_B on the basic form.

```
> fB = function(pH,pKa) {10^(pH-pKa)/(1 + 10^(pH-pKa))}
> qNet = function(qB,pH,pKa) {qB + 1/(1 + 10^(pH-pKa))}
```

We then use these functions to calculate and plot the base fraction and net charge on acetic acid ($pK_a = 4.8$, $q_B = -1$) as a function of pH.

```
> par(mfcol=c(1,2))
> curve(fB(x,4.8),2,7, xlab="pH",main="Fraction Base")
> # Note that the variable must be called x, not pH
> text(3,.8,"pKa = 4.8") # pKa for acetic acid
> curve(qNet(-1,x,4.8),2,7, xlab="pH", ylab="qNet",
> main="Net Charge")
> text(6,-0.2,"qB = -1") # qB for acetic acid
> par(mfcol=c(1,1))
```

As a slightly more complicated example, let's do the calculation of net charge as a function of pH for glutamic acid, which has an α-COOH group with $pK_a = 2.2$ and $q_B = -1$, an α-NH$_3^+$ group with $pK_a = 9.7$ and $q_B = 0$, and a COOH side chain with $pK_a = 4.3$ and $q_B = -1$.

```
> curve(qNet(-1,x,2.2) + qNet(0,x,9.7) + qNet(-1,x,4.3),
1,12, xlab="pH", ylab="qNet",
main="Titration of Glutamic Acid")
> abline(0,0,lty=3)
```

The isoelectric point ($q_{Net} = 0$) appears to be somewhat above 3. We'll see below (Section 3.3.2) how to solve for it directly.

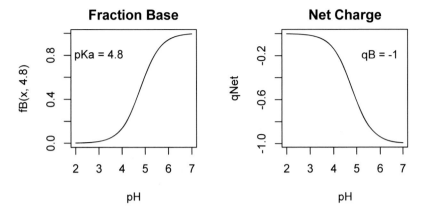

Fig. 3.4 pH titration curves of acetic acid

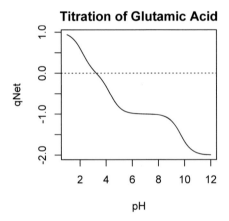

Fig. 3.5 pH titration curves of glutamic acid

3.3 Programming

Programs in R, as in most computer languages, typically consist of a few standard types of operations. We have already encountered two of those operations: assigning the values of variables and evaluating expressions involving those variables. In this section we will introduce two more types of operations: conditional execution, in which different sequences of statements are executed depending on whether an expression is true or false; and repetition or looping, in which an action is performed repeatedly until some condition is met. In the next chapter we'll deal with two other fundamental operations: getting data in and sending data out.

3.3.1 Conditional execution with *if ()*

A key concept in programming is to change the code to be executed depending on some condition(s). In R, this concept is implemented with if statements of the form

```
> if (expr_1) expr_2 else expr_3
```

where expr_1 must evaluate to either TRUE or FALSE. A simple example:

```
> k = 3
> if (k == 3) print("Yes") else print("No")
[1] "Yes"
```

Note that the "logical equals" is "==". Other logical operators used in if statements are & (AND), | (OR), and ! (NOT).

The if...else construct can be applied to an entire vector at once, with the ifelse function, which has the form ifelse(condition, A, B). If condition is true for the ith component of the vector, A[i] is returned; if false, B[i]. For example,

```
> kvec = runif(5,0,2)
> kvec
[1] 0.2260 0.4038 1.6850 1.2795 0.0794
> ifelse (kvec > 1, "T", "F")
[1] "F" "F" "T" "T" "F"
```

3.3.2 Looping with *for ()*

In a for () loop, the program does a repetitive calculation, stepping through the designated values of a running index. As a simple example, consider the countdown to blastoff:

```
> for (i in 3:1) print(i)
[1] 3
[1] 2
[1] 1
```

We can redo the q_{Net} in the section above using a for () loop, in which we calculate the value of the charge for each titratable group, and sum the result.

```
> qNet = function(pH, qB, pKa, n) {
    x=0;
    for(i in 1:length(qB)) {
    x = x+n[i]*qB[i] + n[i]/(1 + 10^(pH - pKa[i]))}
    return(x)
    }
```

where we anticipate generalization of the calculation by allowing for variable numbers *n* of each titratable group. Then

```
> qB=c(-1,0,-1)
> pKa=c(2.2,  9.7,  4.3)
> n=c(1,1,1)
> curve(qNet(x,qB,pKa,n),1,12, xlab="pH", ylab="qNet",
> main="Titration of Glutamic Acid")
```

gives the same graph as above.

Introduction to R points out that "for() loops are used in R code much less often than in compiled languages. Code that takes a 'whole object' view is likely to be both clearer and faster in R." For example, compare

```
> for (i in 0:10){
x[i] = i
y[i] = sin(2*pi*x[i]/10)
}
```

with

```
> x = 0:10
> y = sin(2*pi*x/10)
```

3.3.3 Looping with `while()`

The while() function behaves like the for() function, looping over a series of values, but is used instead of for() when we don't know what values to loop over, but instead keep looping while a condition is met. Here is an example that counts how many tails occur in a series of coin tosses before the first head.

```
> heads.or.tails = function() {
  coin = "tails"     # Initialize
  count = -1          # Get counting right this way
  while(coin == "tails") {
    coin = sample(c("heads","tails"),size=1,replace=T,
       prob=c(.5,.5))
    count=count + 1
    print(coin)
  }
  cat("Found",count,"tails before the first head\n")
}
```

In the last line, "cat" concatenates pieces into a text string and "n" inserts a new line. Run the function in R by typing heads.or.tails().

3.3.4 Looping with `repeat`

`repeat` will loop over an expression until told to `break`. For example:

```
> i = 0
> repeat {print(i); i = i+1; if (i > 5) break}
[1] 0
[1] 1
[1] 2
[1] 3
[1] 4
[1] 5
```

Note that `break` is the only way to exit a `repeat` loop.

3.3.5 Choosing with `which()`

The function `which` gives the indices of a logical object for which the condition is TRUE. Consider, for example, the amino acids with charged side chains:

```
> ChargedAAs=c("Asp","Glu","His","Cys","Tyr","Lys",
  "Arg")
> which(ChargedAAs == "His")
[1] 3
```

We can use the found index to look up properties of the object that meets the desired criterion.

```
> pKa=c(3.9, 4.3, 6.0, 8.3, 10.9, 10.8, 12.5)
> pKa[which(ChargedAAs == "His")]
[1] 6
```

3.4 Numerical analysis with R

R is generally thought of as a programming language for statisticians, but it has the capabilities needed for the sort of numerical analysis done in most sorts of scientific work. In this section we show some of the most common examples: finding the roots of polynomials or other functions, solving systems of linear and nonlinear equations, and numerical integration and differentiation. Other important functions, such as numerically solving differential equations, fitting data to linear or nonlinear equations, and finding periodicities in data with spectral analysis and Fourier transforms, will be introduced in subsequent chapters.

3.4.1 Finding a zero of a function

We often want to find one or more roots of a function, that is, the places at which the function equals zero. This task is carried out in R by the `uniroot` function, which searches a specified interval from `lower` to `upper` for a root. Note that `uniroot`, as its name implies, finds only one root at a time. The function `polyroot`, discussed in the next subsection, finds all the roots, real and complex, of a polynomial; but it does so only for polynomial functions, while `uniroot` works with any kind of function.

We'll show in this section how to use the `uniroot` function to solve for the pH at which $q_{Net} = 0$; that is, the isoelectric point. We define a function f of x, which becomes 0 at some value of x, and we want to know that value. In this case, x is the pH, and the function is q_{Net}. We then call `uniroot(f, lower, upper)` where lower and upper define the range within we think the root lies. (It's often useful to graph the function to get an idea of that range.)

```
> f=function(pH) {qNet(pH,qB,pKa,n)}
> str(uniroot(f,c(2,5)))
List of 4
 $ root       : num 3.25
 $ f.root     : num -4.8e-06
 $ iter       : int 4
 $ estim.prec : num 6.1e-05
```

`str` stands for "string"; it isn't required but gives a neater output. There are four components of uniroot: the root itself, the value of f at the root (should be near 0), how many iterations it took to converge to the root within an acceptable tolerance, and the estimated precision of the result. See `uniroot` in the R help system for more details.

We can generalize this calculation for a polypeptide or protein of arbitrary composition by listing in q_B and pK_a the parameters for the alpha-carboxy and alpha-amino residues and the side chains of all the ionizable amino acids (of which there are seven).

```
> # pKas and qBs for amino acids
> pKa=c(2.2, 9.6, 3.9, 4.3, 6.0, 8.3, 10.9, 10.8, 12.5)
> qB =c(-1, 0, -1, -1, 0, -1, -1, 0, 0)
> names=c("aCOOH","aNH3","Asp","Glu","His","Cys","Tyr",
"Lys","Arg")
```

We then let n be the vector of the numbers of each titratable residue in the molecule. For Glu, `n=c(1,1,0,1,0,0,0,0,0)`. For the hypothetical peptide LysLysGluHisLysLys, `n=c(1,1,0,1,1,0,0,4,0)`, its charge titration curve is shown in Figure 3.6, and the isoelectric point, which appears to be between 10 and 12, is calculated to be 10.8.

```
> str(uniroot(f,c(10,12)))
```

Titration of LysLysGluHisLysLys

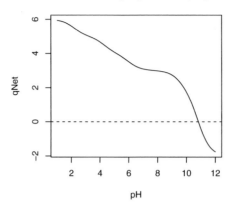

Fig. 3.6 Titration curve of a hexapeptide

```
List of 4
 $ root       : num 10.8
 $ f.root     : num -1.25e-06
 $ iter       : int 5
 $ estim.prec: num 6.1e-05
```

3.4.2 Finding the roots of a polynomial

Although it's relatively easy to find the roots of a linear or quadratic equation with paper and pencil, higher order equations present more of a challenge. R has a useful function, polyroot, for finding all the roots—real and complex—of a polynomial. Consider the function

$$y = x^4 + x^3 - 3x^2 - x + 1 \tag{3.2}$$

We want to find the values of x for which $y = 0$. It's usually a good idea to plot a function, to get an idea of its general behavior and where the roots may lie:

```
> curve(x^4+x^3-3*x^2-x+1,-3,3)
> abline(0,0)
```

It looks as if the roots are near -2, -1, 0.5, and 1.5. To get precise values, use polyroot with the coefficients given in increasing order.

```
> polyroot(c(1,-1,-3,1,1))
[1]  0.4773+0i -0.7376-0i -2.0953+0i  1.3557-0i
```

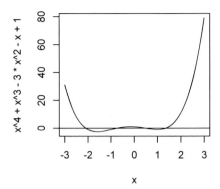

Fig. 3.7 Graphical estimation of the roots of a polynomial

Even though all the roots are real, the imaginary parts, all of which are 0, are also given. Note that, in contrast to `uniroot`, you don't have to enter guesses for the roots.

3.4.3 Solving a system of simultaneous linear equations

Another common situation is having to solve a set of simultaneous linear equations. This is done using matrix manipulations.

An example is to calculate the amounts of different kinds of foods to feed lab animals to be sure that they get the necessary amounts of nutrients. Suppose you have some mice that each day need 200 units of nutrient A, 100 units of B, and 200 units of C. You have three foods: A gram of f_1 has 20 units of A, 18 of B, and 12 of C. A gram of f_2 has 16 units of A, 3 of B, and 0 of C. A gram of f_3 has 2 units of A, 7 of B, and 27 of C. How many grams of each food must you give the mice each day? We have three equations in three unknowns:

$$20f_1 + 16f_2 + 2f_3 = 200$$
$$18f_1 + 3f_2 + 7f_3 = 100 \tag{3.3}$$
$$12f_1 + 27f_3 = 200$$

This can be put in matrix form as follows:

$$\begin{pmatrix} 20 & 16 & 2 \\ 18 & 3 & 7 \\ 12 & 0 & 27 \end{pmatrix} \begin{pmatrix} f_1 \\ f_2 \\ f_3 \end{pmatrix} = \begin{pmatrix} 200 \\ 100 \\ 200 \end{pmatrix} \tag{3.4}$$

The general solution of such a matrix equation is

$$\mathbf{M}f = y \rightarrow f = \mathbf{M}^{-1}y \qquad (3.5)$$

where $\mathbf{M}^{-1}$ is the inverse of the matrix $\mathbf{M}$. In R, one calculates the inverse of a matrix using `solve()`, so we can calculate the amounts of each feedstock as follows:

```
> M=matrix(c(20,16,2,18,3,7,12,0,27),nrow=3,ncol=3,
  byrow=T)
> M
      [,1] [,2] [,3]
[1,]   20   16    2
[2,]   18    3    7
[3,]   12    0   27
> Minv=solve(M)
> Minv
             [,1]        [,2]        [,3]
[1,] -0.016585   0.08845  -0.02170
[2,]  0.082310  -0.10565   0.02129
[3,]  0.007371  -0.03931   0.04668
> y=c(200,100,200)
> f=Minv%*%y
> f
          [,1]
[1,]   1.188
[2,]  10.156
[3,]   6.880
```

Observe the operator notation `%*%` for matrix multiplication, or multiplication of a vector by a matrix.

3.4.4 Solving a system of nonlinear equations

We can solve a set of simultaneous nonlinear equations with the `optim()` function, which by default minimizes a function. Consider the set of equations

$$x_1^2 + x_2^2 = 1 \qquad (3.6)$$

$$\sin(\pi x_1/2) + x_2^3 = 0 \qquad (3.7)$$

that we want to solve for x_1 and x_2. We subtract 1 from both sides of the first equation, so we have two functions that are equal to zero. If we square them, they are still equal to zero, and if we add the squares the sum should be equal to zero, with the proper values of x_1 and x_2. To find those values we optimize (minimize) the function that equals the sum of squares, as follows:

```
> fr=function(x) {
```

```
x1=x[1]; x2=x[2]
# Return the function to be optimized
(x1^2+x2^2-1)^2 + (sin(pi*x1/2)+x2^3)^2
}
> # Guess that the values are between +/- 1
> optim(c(-1,1),fr)
$par
[1] -0.4760  0.8794

$value
[1] 7.45e-09

$counts
function gradient
      67      NA

$convergence
[1] 0

$message
NULL
```

optim returns not only the roots, but also other information about the solution
process. To get only the roots, you can do the following:

```
> roots = optim(c(-1,1),fr)$par
> roots
[1] -0.4760  0.8794
```

Note that fr is a function of the vector variable x, which has two components.
See optim() in the R help system for details on how this works, and on the variations available. Note also that if your initial guess is c(1, -1) for the two roots,
you'll get roots of opposite sign. Try it.

3.4.5 Numerical integration of functions

R has the function integrate for numerical integration of functions of one variable. For example, to integrate the sine function between 0 and π:

```
> integrate(sin,0,pi)
2 with absolute error < 2.2e-14
```

More complicated functions must be defined explicitly:

```
> f = function(x) sin(x)^2 * exp(-x/5)
> integrate(f,0,10)
2.108 with absolute error < 1.4e-12
```

3.4.6 Numerical integration of data using splinefun

Sometimes functions cannot be expressed or integrated analytically, or you are confronted with some experimental data that needs to be integrated (e.g., the area under an HPLC curve to determine the amount of material in a protein band). In such cases you need to call on numerical integration methods. Texts on numerical analysis present a variety of methods, typically progressing from rectangular through trapezoidal to Simpson's rule integration and sometimes beyond. However, the simplest way to do numerical integration of data in R is to combine integrate with splinefun. For example, suppose we have only some (x,y) points defining the sine function.

```
> x = seq(0, pi, length = 10)
> y = sin(x)
```

Then we fit those points to a spline function and integrate the function over the desired range.

```
> f = splinefun(x,y)
> integrate(f,0,pi)
2 with absolute error < 7.5e-05
```

The precision is lower, but the result is the same as for integration of the analytic function, above.

3.4.7 Numerical differentiation

Numerical differentiation of experimental curves is often useful to emphasize minima, maxima, and inflection points. Once the spline function fit to a set of (x,y) points has been obtained, it can be used for numerical differentiation to obtain first, second, and third derivatives. We again use the sine function for simplicity.

```
> plot(x,y, ylim = c(-1,1)) # x, sin(x) points
> curve(f(x), add = T) # Spline through points
> # 1st deriv
> curve(f(x, deriv=1), add = T, lty="dashed")
> # 2nd deriv
> curve(f(x, deriv=2), add = T, lty="dotted")
> abline(0,0) # Show zero
```

This procedure using splines will be adequate for many purposes. If more precision is needed, the numDeriv add-in package will be useful. The base installation of R also has a function diff() that can be used, with suitable adjustment of arguments, to calculate first- and higher-order differences in a vector or matrix of numerical values.

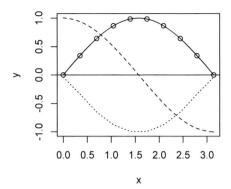

Fig. 3.8 First- and second-order numerical differentiation using splines, along with the original function

3.5 Problems

1. The Lorentzian function often appears in spectroscopy. It is described by the equation

$$L(x) = \frac{1}{\pi} \frac{w}{(x - x_0)^2 + w^2} \qquad (3.8)$$

where w is the half-width at half-height (the x-distance from the maximum at x_0 to where the function has half its maximum value). As written, the function is normalized to unity: the total area under the curve = 1. The Lorentzian function is superficially similar to the Gaussian, but has wider tails. Write an R function for the Lorentzian, and use it to plot the Lorentzian and Gaussian on the same graph, with $x_0 = 10$ and w or $sig = 2$ in both cases.
2. Plot a spectrum with two Lorentzians, one of magnitude 2, the other of magnitude -1 (negative), with $x_0 = 270$ and 250, respectively, and $w = 10$ for both. Run the plot from $x = 200$ to 320.
3. Write a function to count the number of bases of each type (A, C, G, T) in a sequence, and apply it to a randomly generated sequence of 200 bases in which each appears with equal probability. You should use the `for` statement to loop over each base in the sequence, and the `if` statement to test for each of the bases. Set a counter = 0 for each of the bases initially, and add 1 to it each time the `if` encounters a base of that type. Return a vector that gives the number of each type.
4. Use `while` to implement Newton's method to find the zeros of a function. The interative steps in Newton's method are

$$x_{n+1} = x_n - \frac{f(x_n)}{f'(x_n)} \qquad (3.9)$$

For example, to find the zeros of $f(x) = x^2 - \sin(x)$, we iterate the equation

$$x_{n+1} = x_n - \frac{x_n^2 - \sin(x)}{2x_n - \cos(x_n)} \qquad (3.10)$$

until the difference between x_n and x_{n+1} is less than some small value delta. Pseudocode for this process can be written as

```
while(delta > .00001) {
    old.x = x
    x = x - (x^2 - sin(x))/(2*x - cos(x))
    delta = | x - old.x |
}
```

Write and run an R function to implement Newton's method for the function given, finding both roots. Note that your answer will depend on your starting value of x, and you are advised to graph the function to get a first estimate of the roots.

5. Sometimes functions cannot be integrated analytically, or you are confronted with some experimental data that needs to be integrated (e.g., the area under an HPLC curve to determine the amount of material under a peak). In such cases you need to call on numerical integration methods. The simplest method is to lay out n equally spaced points along the x-axis between lower limit $x_1 = a$ and upper limit $x_n = b$. This divides the x-axis into intervals of width

$$\Delta x = \frac{x_n - x_1}{n - 1} \qquad (3.11)$$

and we want to do the integral

$$I = \int_{x_1}^{x_n} f(x)dx \qquad (3.12)$$

by summing $f(x)$ over the $n - 1$ intervals of width Δx.
The most elementary way to proceed is via "rectangular integration", by assuming that the value of $y = f(x)$ is constant over each interval. This gives a "zeroth-order" approximation in which the curve is turned into a "staircase" which underestimates the areas as the curve goes up and overestimates as the curve goes down:

$$I = (y_1 + y_2 + y_3 + \ldots + y_{n-1})\Delta x \qquad (3.13)$$

Write and run an R program to integrate $f(x) = 1 + x^2$ over the interval from 0 to 2, using the rectangular approximation. What value of n do you need in order to get a value for the integral within 0.1% of the true value of 4 2/3 = 4.667?

6. The first-order approximation for numerical integration is called "trapezoidal integration", in which successive y points are connected by straight lines, giving a series of trapezoids whose areas are added. It can easily be shown that this differs from the rectangular approximation only by a term involving the end-points of the curve: $I_{trap} = I_{rect} + 0.5(y_n - y_1)\Delta x$.

Write and run an R program to integrate $f(x) = 1 + x^2$ over the interval from 0 to 2, using the trapezoidal approximation. What value of n do you need in order to get a value for the integral within 0.1% of the true value?

Then use the same program to calculate the area A of a circle of radius $r = 2$, using the equation for a circle $x^2 + y^2 = r^2$ to get y as a function of x and dividing the circle into a series of vertical strips. What value of n do you need in order to get within 0.1% of the exact value $A = \pi r^2$? This approach would be useful, for example, in calculating the area of a region in a micrograph.

7. A better numerical integration method for more rapidly varying functions is known as Simpson's rule, which is based on fitting successive points to a 2nd-order polynomial. The equation for Simpson's rule integration is

$$I = \frac{1}{3}(y_1 + 4y_2 + 2y_3 + 4y_4 \ldots + 2y_{n-2} + 4y_{n-1} + y_n)\Delta x \qquad (3.14)$$

Use Simpson's rule with $n = 20$ to calculate the areas under the Gaussian and Lorentzian curves from $-5 * sig$ to $+5 * sig$ or $-5w$ to $+5w$ on either side of x_0, to show that the curves are indeed normalized to unity.

8. Numerical differentiation of experimental curves is often useful to emphasize minima, maxima, and inflection points. The numerical derivative of a function at the location half-way between points i and $i + 1$ is

$$\left.\frac{dy}{dx}\right|_{i+1/2} = \frac{y_{i+1} - y_i}{h} \qquad (3.15)$$

where $h = \Delta x$ in our previous notation. Write and run an R program to calculate and plot the numerical derivative curve of the spectrum given as a sum of three Gaussians at the beginning of this module. Let $h = 1$ nm in your calculations.

9. Data are often noisy, and the underlying behavior is often better seen if the noise is smoothed by averaging in a "moving window". To do this, an odd number k of consecutive points is chosen, and the smoothed point at the middle is the average of the points in the window. For example, if the window has width 11, then the smoothed value of point 100 is taken as the average of points 95 to 105. If there are n total points, the first smoothed point will be at $(k+1)/2$, and the last at $n - (k-1)/2$.

Apply 10% random noise (rnorm with mean = 0 and sd = 0.1) to the three-Gaussian spectrum at the beginning of this module, with one point every 1/2 nm between 150 nm and 350 nm. Then write and run an R program to smooth the noisy spectrum with windows of 7, 15, and 21 points. For each of the three window sizes, plot the raw data, the smoothed data, and the underlying sum of Gaussians, on three appropriately labeled graphs stacked vertically (using par(mfcol=c(3,1))). Use different colors or line types to differentiate the raw, smoothed, and underlying curves.

10. Apply numerical differentiation (Problem 8) to the data in Problem 9 smoothed with a 15-point window. Comment on how the results (zero crossings, maxima and minima) compare with the results in Problem 8.

Chapter 4
Data and Packages

Up to this point we have mainly dealt with how to use R for calculating and graphing. In programming for scientific work we also generally need to get data from various sources, transform it, and save it for later use. We also will often wish to augment the built-in capabilities of R with more specialized resources. Many such resources are available as contributed packages from the CRAN web site. This chapter will deal with those two important topics: handling data and adding packages.

4.1 Writing and reading data to files

If you are doing substantial work with a set of data, you will want to be able to save it as a file, and to recall it when needed. You may also get data from other sources, for analysis or manipulation.

4.1.1 Changing directories

By default, R puts your files in your home directory, which it considers the "working directory". You may want to change the working directory for some purpose. You can find out what the current directory is by using the function getwd(). If I apply that function on my Mac, I get

```
> getwd()
[1] "/Users/victor"
```

To change the working directory, use setwd. For example, to change the working directory to the Desktop within my home directory,

```
> setwd("/Users/victor/Desktop/")
```

or more concisely, since ~ represents the user's home directory

V. Bloomfield, *Computer Simulation and Data Analysis in Molecular Biology and Biophysics,*
Biological and Medical Physics, Biomedical Engineering, DOI: 10.1007/978-1-4419-0083-8_4,
© Springer Science + Business Media, LLC 2009

```
> setwd("~/Desktop/")
> getwd()  # Check whether it worked
[1] "/Users/victor/Desktop"  # Mission accomplished.
```

Alternatively, to stay in the default directory, but save or load a file from another directory:

```
> setwd("~")  # Set working directory back to default
> getwd()  # Check that I'm back where I want to be
[1] "/Users/victor"
```

Then to import the file my.file from the Desktop using the scan function:
```
scan(file="/Users/victor/Desktop/my.file")
```

4.1.2 Writing data to a file

Suppose you have generated the following simulated data:

```
> x=1:5
> y1=log(x)
> y2=sin(x)
> y3=rnorm(length(x))
> cxy=cbind(x,y1,y2,y3)
> cxy
        x      y1        y2         y3
[1,]  1 0.0000   0.8415   0.52823
[2,]  2 0.6931   0.9093  -0.04237
[3,]  3 1.0986   0.1411   0.78224
[4,]  4 1.3863  -0.7568   1.98371
[5,]  5 1.6094  -0.9589   0.07465
```

cbind binds vectors into a matrix by columns; rbind does so by rows.

The command to write an array of data my.data to a plain text file is write(my.data, file="my.file"). So we write

```
> write(cxy,file="cxy")
```

4.1.3 Reading data from a file using scan()

We can read the data back from the saved file in the working directory using scan():

```
> scan("cxy")
Read 20 items
 [1]   1.00000   2.00000   3.00000   4.00000   5.00000
```

```
        0.00000   0.69310   1.09900
  [9]   1.38600   1.60900   0.84150   0.90930   0.14110
       -0.75680  -0.95900   0.52820
 [17]  -0.04237   0.78220   1.98400   0.07465
```

but we see that this doesn't preserve the dimensions of the matrix. You would have to re-dimension it by `matrix(nrows=..., ncols=..., byrows = T or F)}`. Also, `scan()` reads only numbers, not text.

4.1.4 Writing and reading tables

A better way to save the table with formatting and headings is to use the commands `write.table` and `read.table`.

```
> write.table(cxy,file="cxy")
> cxy=read.table(file="cxy", header=T)
```

Of course, you may give the file another name when writing, and the table another name when reading it in.

4.1.5 Reading and writing spreadsheet files

If you have data in a spreadsheet and want to analyze it with R, save it as a plain text file (.txt) or a comma-separated (.csv) file. If the file is saved in tab-separated format, then one retrieves it using

```
> read.table(file="/Users/victor/Desktop/sheet.txt",sep="\t")
```

where `\t` is the symbol for the tab character. If the file is saved in csv format, then use

```
> read.table(file="/Users/victor/Desktop/sheet.csv",sep=",")
```

To write a file that can be imported by a spreadsheet, simply use `write.table`.

4.1.6 Saving the R environment between sessions

It is possible to save an R object, or environment, between sessions simply by `save()`. (See R Help.) However, this is generally not the best way to preserve data objects in usable form. See Data Frames (Section 4.3), to see how to do this properly in R.

We have just touched the basics of data import and export. For a deeper view, the *R Data Import/Export* document in R Help describes the import and export facilities available either in R itself or via packages which are available from CRAN.

4.2 Packages

The installation of R on your computer contains a library of packages. A package may contain data sets, functions written in R, and dynamically loaded libraries of C or Fortran code. Some packages are part of the basic R installation, others can be downloaded from CRAN or through the Package Manager/Installer in your R program, and some you might write yourself. Most packages will not be needed all the time, so they are loaded into R using the `library('package.name')` command. To find what packages are currently installed in R on your computer, type `library()`.

One of the packages in the base installation is `datasets`. To see what data are available in that package, type `data(package="datasets")`. The first few data sets are

```
Data sets in package 'datasets':

AirPassengers            Monthly Airline Passenger Numbers
                         1949-1960
BJsales                  Sales Data with Leading Indicator
BJsales.lead (BJsales)
                         Sales Data with Leading Indicator
BOD                      Biochemical Oxygen Demand
CO2                      Carbon Dioxide uptake in grass plants
ChickWeight              Weight versus age of chicks on different
                         diets
DNase                    Elisa assay of DNase
EuStockMarkets           Daily Closing Prices of Major European
                         Stock Indices, 1991-1998
Formaldehyde             Determination of Formaldehyde
```

Use `data(package = .packages(all.available = TRUE))` to list the data sets in all *available* packages. Try it to see what you get.

To actually load a package, type `library(package.name)`. To find out what is in a particular data set in a library that you've loaded, type a question mark followed by the name of the data set. For example:

```
> library(datasets)
> ?BOD
BOD {datasets}      R Documentation
Biochemical Oxygen Demand
Description
The BOD data frame has 6 rows and 2 columns giving the
biochemical oxygen demand versus time in an evaluation
of water quality.
```

This is followed by sections on Usage, Format (of the variables), Source (of the data), and Examples (code showing how the data might be used). Try it for yourself with several of the data sets in `datasets`.

The variables in the data set `Puromycin`, for example, are `conc`, `rate`, and `state`. To access these variables you would have to type `Puromycin$conc`, `Puromycin$rate`, etc. To avoid the rather clumsy `$` notation, you can use `attach()`. Then the variable names themselves can be used:

```
> attach(Puromycin)
> conc
 [1]  0.02 0.02 0.06 0.06 0.11 0.11 0.22 0.22 0.56 0.56
      1.10 1.10 0.02 0.02 0.06
[16]  0.06 0.11 0.11 0.22 0.22 0.56 0.56 1.10
> rate
 [1]  76   47   97 107 123 139 159 152 191 201 207 200
      67   51   84   86   98 115 131
[20] 124 144 158 160
```

You can then do things like `plot(conc, rate)`.

Note, however, that the `DNase` data set also has a `conc` variable, so if you now type `attach(DNase)` you get the warning message

```
The following object(s) are masked from Puromycin :

  conc
```

That is, the `DNase` variables are put "ahead" of the `Puromycin` variables in the R search path. (Type `search()` to see how things are arranged.) To make the `conc` variable in `Puromycin` accessible without using the construction `Puromycin$conc`, you have to type `detach(DNase)`. Try it and see.

4.3 Data frames

Scientists typically arrange data in tables, usually labeled at the top by column names to indicate variables, and perhaps by row names to indicate each measured sample. (The analog is a database table, in which the columns are variables and the rows are records.) In R, the corresponding object is the "data frame". Dalgaard [15, p. 18] writes that a data frame "is a list of vectors and/or factors of the same length, which are related 'across', such that data in the same position come from the same experimental unit (subject, animal, etc.). In addition it has a unique set of row names."

Consider the `BOD` dataframe from `datasets`.

```
> BOD
  Time demand
1    1    8.3
2    2   10.3
3    3   19.0
4    4   16.0
```

```
5     5     15.6
6     7     19.8
```

The individual variables can be accessed using the $ notation:

```
> BOD$Time
[1] 1 2 3 4 5 7
> BOD$demand
[1]   8.3 10.3 19.0 16.0 15.6 19.8
```

Just as we can access individual elements of a vector by giving its indices vec[2], vec[c(1,3,5)], vec[2:4], etc., we can access the elements of a data frame:

```
> BOD[4,2] # Column 2 (demand) for Time 4
[1] 16
> BOD[4,]
   Time demand # All columns for Time 4
4     4      16
```

To extract the data for all cases that meet some criterion, e.g., demand >= 16

```
> BOD[BOD$demand >= 16,]
   Time demand
3     3    19.0
4     4    16.0
6     7    19.8
```

To name or rename the rows of a data frame, use rownames(my_frame). To do the same for the columns, use names(my_frame). For example,

```
> rownames(BOD)=c("Mon","Tue","Wed","Thu","Fri","Sun")
> names(BOD) = c("Day",  "Demand")
> BOD
       Day Demand
Mon     1     8.3
Tue     2    10.3
Wed     3    19.0
Thu     4    16.0
Fri     5    15.6
Sun     7    19.8
```

Data frames can be expanded with additional information. For example, we construct a data frame consisting of protein names, molecular weights, and specific volumes.

```
> Protein=c("Lipase","Cytochrome C","Metallothionein")
> MolWt = c(6669, 13370, 9720)
> SpVol = c(0.714, 0.728, 0.648)
```

We form the data frame ProteinData by combining these three vectors by column.

```
> ProteinData = data.frame(Protein, MolWt, SpVol)
> ProteinData
          Protein MolWt SpVol
1          Lipase  6669 0.714
2     Cytochrome C 13370 0.728
3 Metallothionein  9720 0.648
```

Now suppose we want to tabulate the molar volume (mL/mol) of the proteins, multiplying their specific volumes (mL/g) by their molecular weights (g/mol).

```
> MolVol = MolWt*SpVol
> MolVol
[1] 4761.666 9733.360 6298.560
```

There are too many significant figures, so we round.

```
> MolVol = round(MolVol,0)
> MolVol
[1] 4762 9733 6299
```

Finally, we incorporate the MolVol vector into the data frame.

```
> ProteinData = cbind(ProteinData,MolVol)
> ProteinData
          Protein MolWt SpVol MolVol
1          Lipase  6669 0.714   4762
2     Cytochrome C 13370 0.728   9733
3 Metallothionein  9720 0.648   6299
```

4.4 Factors

Factors in R are categorical variables that indicate subdivisions of the data set. For example, in the data frame DNase, the factor Run is a vector of numbers that indicate the assay run; while in Puromycin the factor state is a vector of two strings, treated and untreated, indicating the state of treatment of the cells. The individual values of the factor are called "levels" of which there are eleven for DNase and two for Puromycin. To encode a vector of levels as a factor, use the factor function.

The function tapply can be used to apply a function to a table subdivided by an index, which in this case will be a factor. Consider the data frame InsectSprays in datasets, which tabulates "[t]he counts of insects in agricultural experimental units treated with different insecticides". It has two columns: count which gives the numbers of insects, and spray, a factor of letters A–F that indicates the type of spray. Then we can get the mean number of insects for each spray type by

```
> tapply(InsectSprays$count, InsectSprays$spray, mean)
      A       B       C       D       E       F
14.500  15.333   2.083   4.917   3.500  16.667
```

A data frame may be split into sub-tables according to its factors with the `split` function, e.g., `split(InsectSprays, InsectSprays$spray)`. Alternatively, a particular level may be captured like this:

```
> InsectSprays.A = InsectSprays[InsectSprays$spray == "A",1]
> InsectSprays.A
 [1] 10  7 20 14 14 12 10 23 17 20 14 13
```

4.5 The contributed package `seqinr`

R contains numerous contributed packages that are useful for molecular biology and biophysics. Here we shall focus on one, `seqinr`, which contains a remarkable assortment of tools for analyzing protein and nucleic acid sequences. If it's not already on your machine, download the package from the CRAN web site, load it with

```
library(seqinr)
```

and then get information about its contents with

```
library(help=seqinr)
```

This is a very large package, so the list of available functions and data will take a while to appear fully on your screen.

For example, look at `AAstat` to see what protein statistics it yields:

```
?AAstat
```

We see that it requires the amino acid sequence of the protein in one-letter, upper-case notation. It returns the count of each amino acid in the protein, a list giving the percentage of each physico-chemical class, and the computed isoelectric point. If desired, it will also plot the distribution of physico-chemical classes along the sequence.

In the first few functions we also see a, which converts a three-letter amino acid code into a one-letter code, and `aaa`, which does the reverse.

A particularly remarkable component of `seqinr` is aaindex, which lists 544(!) physico-chemical and biological properties for the 20 amino acids, taken from the work of Kanehisa and collaborators (Kawashima, S. and Kanehisa, M. (2000) AAindex: amino acid index database. *Nucleic Acids Res.*, 28:374). Load the list with `data(aaindex)` and get information about it with `?aaindex`. Scrolling down the long list of properties, we note that it will be very difficult to find and use a desired property unless we know what we're looking for. Fortunately, the `Examples` section of aaindex tells us how to do that.

```
aaindex is formatted as follows:
```

```
A named list with 544 elements having each the
following components:
H
String: Accession number in the aaindex database.
D
String: Data description.
R
String: LITDB entry number.
A
String: Author(s).
T
String: Title of the article.
J
String: Journal reference and comments.
C
String: Accession numbers of similar entries with the
correlation coefficients of 0.8 (-0.8) or more (less).
Notice: The correlation coefficient is calculated with
zeros filled for missing values.
I
Numeric named vector: amino acid index data.
```

Therefore, the data we want is in field aaindex$I, and we can search for an author of the paper that generated the data in {aaindex$A}.

If we want the Kyte-Doolittle hydropathy index, we first look at the entries with Kyte as author:

```
> which(sapply(aaindex, function(x)
   length(grep("Kyte", x$A)) != 0))
KYTJ820101
       151
```

Let's deconstruct that which statement. which(v) returns a vector consisting of those elements of v that are TRUE. v in this case is the result of applying a function to all elements of the sequence aaindex with sapply. The function asks for those entries in x for which the length of a search (grep) for "Kyte" in the Author ($A) field of x is not equal to zero (!=0). If the length is zero, FALSE is returned. Only one entry returns TRUE. Single elements of lists are picked out by double square brackets: [[...]]. The field aaindex[[151]]$I of entry 151 then gives the desired hydropathy values in the $I field.

```
> aaindex[[151]]$I
Ala  Arg  Asn  Asp  Cys  Gln  Glu  Gly  His  Ile  Leu
1.8 -4.5 -3.5 -3.5  2.5 -3.5 -3.5 -0.4 -3.2  4.5  3.8
 Lys  Met  Phe  Pro  Ser  Thr  Trp  Tyr  Val
-3.9  1.9  2.8 -1.6 -0.8 -0.7 -0.9 -1.3  4.2
```

`aaindex[["KYTJ820101"]]$I` and `aaindex$KYTJ820101$I` give the same results.

Note that `aaindex` requires the three-letter amino acid code.

Now let's use the Kyte-Doolittle hydropathy index to determine the distribution of hydropathy along the sequence of a protein. Go to the Entrez web site, `http://www.ncbi.nlm.nih.gov/sites/gquery` choose the `Protein` sequence database, and search `Protein` for `dutpase`. Several thousand hits result, but let's choose the one with index AAA46131. Display FASTA to get the amino acid sequence as a string in one-letter, upper-case code. Copy it from your browser screen and paste it into R as the variable `dutpase`, with the string in quotes.

```
> dutpase <-
"MPYKVPEIYYRFEPQTFYITSPARASNLQLINHNNILVKAGQVTIVSTGIIFP
KETSFAFILYGKSAKSIFCHTGLIDPGFQGELKLIVLNKTEDDITLFENDLRVS
VTAFVYGVPKLHDYSDLCPPRYSKDAGFDLYLPTDVTVKPRVPNRYSVNICCPA
QLKSYKPVLFGRSGLAAKGLTIKVSRWQNQLQIIFYNYTKSQITYTARTRIAQV
VFMHKKHLPTTLTRLKPTMHLSENIKYSWARVSFQDIKTFPVQDEKLYSSSKDT
SDSQMSRGDAGLGSSGLM"
```

To manipulate this amino acid sequence, we need to turn the string into a vector of single characters. This is accomplished using the `seqinr` function `s2c`. (You can also use the basic R function `strsplit`.)

```
> dUTPase1 = s2c(dutpase)
```

If you inspect this sequence, you'll find occasional elements `"\n"` that represent new line feeds rather than amino acids:

```
> dUTPase1
...
 [71] "\n" "F"  "C"  "H"  "T"  "G"  "L"  "I"  "D"  "P"
      "G"  "F"  "Q"  "G"
...
```

These need to be removed from the sequence vector before further manipulation can take place:

```
> dUTPase1 = dUTPase1[dUTPase1 != "\n"]
```

That is, we select for dUTPase1 only those elements that do not contain `"\n"` with the logical operator `|=` (not equal to).

```
> dUTPase1
... [55] "E" "T" "S" "F" "A" "F" "I" "L" "Y" "G" "K"
        "S" "A" "K" "S" "I" "F" "C"
 [73] "H" "T" "G" "L" "I" "D" "P" "G" "F" "Q" "G"
        "E" "L" "K" "L" "I" "V" "L"
...
```

At this point we can determine the molecular weight of the protein with the pmw function of `seqinr`.

```
> pmw(dUTPase1)
[1] 32507.13
```

The sequence can now be converted to a three-letter amino acid code with `seqinr`'s aaa:

```
> dUTPase3
... [61] "Ile" "Leu" "Tyr" "Gly" "Lys" "Ser" "Ala"
         "Lys" "Ser" "Ile" "Phe" "Cys"
  [73] "His" "Thr" "Gly" "Leu" "Ile" "Asp" "Pro" "Gly"
       "Phe" "Gln" "Gly" "Glu"
...
```

If we wish, the reverse conversion from three-letter to one-letter code can be done with a().

We are now ready to determine the hydropathy along the sequence of dUTPase. First we rename the aaindex data to something more comprehensible.

```
> KDHydro = aaindex[[151]]$I;
```

Then we check it out with a short sequence:

```
> seq=c("Ala","Met","Ile")
> KDHydro[seq]
Ala Met Ile
1.8 1.9 4.5
```

Satisfied that all is working properly, we run the function on the full dUTPase sequence:

```
> plot(KDHydro[dUTPase3],type="h")
```

We conclude this introduction to the capabilities of `seqinr` with application of the function AAstat (which uses the one-letter code):

```
> AAstat(dUTPase1,plot=T)

$Compo

  *  A  C  D  E  F  G  H  I  K  L  M  N  P  Q  R  S  T  V  W  Y
  0 14  4 14  8 15 15  6 20 22 26  5 11 17 13 13 25 22 19  2 16

$Prop
$Prop$Tiny
[1] 0.2787456

$Prop$Small
[1] 0.4912892

$Prop$Aliphatic
```

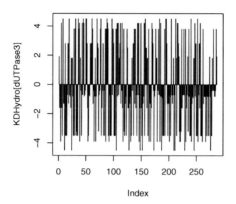

Fig. 4.1 Hydropathy plot for dUTPase, using Kyte-Doolittle parameters

```
[1] 0.2264808

$Prop$Aromatic
[1] 0.1358885

$Prop$Non.polar
[1] 0.533101

$Prop$Polar
[1] 0.466899

$Prop$Charged
[1] 0.2195122

$Prop$Basic
[1] 0.1428571

$Prop$Acidic
[1] 0.07665505

$Pi
[1] 9.488414
```

Note that the isoelectric point is calculated as part of AAstat. If we had wanted only the isoelectric point, we could have used computePI(dUTPase1).

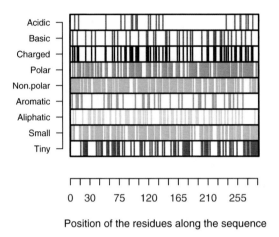

Position of the residues along the sequence

Fig. 4.2 Plots of various amino acid properties along the sequence of dUTPase. On the computer screen the properties will appear in color

4.6 Problems

1. A combination of R Help and the R Project web site gives essentially as complete and useful reference material on the R language as is available from any of the books on the subject, especially when combined with the examples in the R & BioConductor Manual. Use these facilities to describe briefly the relations between vectors, factors, arrays, matrices, lists, and data frames.
2. What are the functions that enable conversion among the various data structures in Problem 1?
3. Establish a directory (folder) called `Rstuff` in your home directory. In a spreadsheet program, construct a table with several rows and columns, including column headings and row labels, and save it as a text file in `Rstuff`. Issue the R command to import the table into the R workspace. Use R to add a new column, which is some arithmetic combination of the existing columns. Save the new table into `Rstuff` with a new name (don't overwrite the existing file). Then import the new table back into R.
4. Construct two vectors of length 10, the first the years between 1998 and 2007, the second a set of 10 randomly distributed random numbers between 0 and 2. Combine the two vectors into a data frame, with suitably chosen names. Extract the years that correspond to random numbers greater than 1.

5. Access the data set "Formaldehyde" in the package "datasets" in your R installation. What is its description? Carry out the example at the end of the documentation. Save the data to your Rstuff folder.

6. Find another data set in datasets that looks interesting, and repeat Problem 5 with it.

7. Use the Package Manager in the R menu to see which packages are installed on your computer, and which of those are loaded. If ISwR from Dalgaard is not installed, use the Package Installer to do so, or go to the CRAN site to download the appropriate binary file. Load the ISwR package, open the "malaria" data set, and do the example at the end. Plot antibody level as a function of age.

8. The MASS package, which comes as part of the R installation, has many data sets as well as functions. Load MASS, get information on the "muscle" data set, and run the Example at the end. You may not yet understand all the commands and functions that are used, but this is a good example of the power and versatility of R.

9. MASS also contains many useful functions. As a simple example, the function area is a moderately sophisticated integration function. Since you have already loaded MASS, area is available to be called. Define a function of x

```
> fnx = function(x)  sin(x)cos(x/2)x^2
```

plot it between 0 and 2, and integrate it using area between the same limits.

10. The Bioconductor project is one of the most prominent and sophisticated uses of R in a biological context. Look up the LIMMA package on the Web, and copy and paste the brief overview given there.

Part II
Simulation of Biological Processes

Chapter 5
Equilibrium and Steady State Calculations

Much of molecular biology and biophysics deals with the equilibrium and dynamics of biochemical reactions. These will be the topics of this chapter and the next. In this chapter we focus on three important types of reactions: ligand binding, helix-coil transitions in oligonucleotides, and steady-state enzyme kinetics. These serve as test beds for showing how to use the plotting and data analysis capabilities of R.

5.1 Calculation of the concentration of species in a reacting mixture at equilibrium

5.1.1 Binding of a ligand L to a protein or polymer P

In Chapter 3 we showed how to use the R function `optim()` to solve systems of nonlinear equations. (See Section 3.4.4.) This is a particularly useful function to calculate the equilibrium concentration of species in a reacting mixture. We'll focus first on the important biological reaction of ligand (L) binding to a protein (P), with dissociation constant K_d. In the simplest case, we have three species—P, L, and PL—whose concentrations (denoted by square brackets) are calculated from two conservation equations

$$[P]_{tot} = [P] + [PL] \tag{5.1}$$
$$[L]_{tot} = [L] + [PL] \tag{5.2}$$

and the equilibrium expression

$$K_d = \frac{[P][L]}{[PL]} \tag{5.3}$$

   A preliminary caution: If you use molar concentrations, with actual concentrations in the range 10^{-6}, the calculation generally will not converge properly. (Try it

V. Bloomfield, *Computer Simulation and Data Analysis in Molecular Biology and Biophysics,*
Biological and Medical Physics, Biomedical Engineering, DOI: 10.1007/978-1-4419-0083-8_5,
© Springer Science + Business Media, LLC 2009

for yourself.) So express all concentrations and dissociation constants in millimolar, micromolar, or nanomolar, depending on the situation. Bearing this caution in mind, we define a function `bind1` that `optim` is to minimize, give the values of the total concentrations and dissociation constant as parameters, and run the optimization.

```
> bind1 = function(x, Pt, Lt, K) {
Lf = x[1]; Pf = x[2]; PL = x[3];
(Pt-Pf-PL)^2 + (Lt-Lf-PL)^2 + (Pf*Lf-K*PL)^2
}

> Ptot = 1; Ltot = 10; Kd = 4

> y = optim(c(.5, 1, .5), bind1, method = "L-BFGS-B",
lower = c(0,0,0), upper = c(Ltot, Ptot, Ptot),
Pt = Ptot, Lt = Ltot, K = Kd)
> y$par # Display just the equilibrium concentrations
[1] 9.3007 0.3007 0.6993
```

Note that we have used the `"L-BFGS-B"` method that constrains lower and upper bounds; see the R help system for details. It's particularly important to set the lower bound greater than or equal to zero, to avoid getting negative concentrations.

If we want to do the calculation for several values of the ligand concentration, we can give `Ltot` as a vector:

```
> Ptot = 1; Ltot = c(5,10,20); Kd = 4
> y = function(i) optim(c(1, .5, .5), bind1, method =
"L-BFGS-B", lower = c(0,0,0), upper = c(Ltot[i], Ptot,
 Ptot), Lt=Ltot[i], Pt=Ptot, K=Kd)
> ypar = function(i) y(i)$par
> # Check for the same result as above Ltot = 10
> ypar(2)
[1] 9.3007 0.3007 0.6993

# Now display the complete set of results in a matrix
> yp = matrix(nrow=length(Ltot), ncol=2+length(Kd),
  byrow=T)
> for (i in 1:length(Ltot)) yp[i,]=y(i)$par
> yp
        [,1]    [,2]    [,3]
[1,]   4.472 0.4721 0.5279
[2,]   9.301 0.3007 0.6993
[3,] 19.173 0.1726 0.8274
```

We can make the matrix more comprehensible with row and column names.

```
> colnames(yp) = c("Lf","Pf","PL")
> rownames(yp) = c("5","10","20")
> yp
```

```
        Lf       Pf       PL
 5    4.472  0.4721  0.5279
10    9.301  0.3007  0.6993
20   19.173  0.1726  0.8274
```

5.1.2 Fitting binding data with a Scatchard plot using `linear model` (`lm`)

The extent of binding is given by $v = [PL]/[P]_{tot}$, the ratio of occupied binding sites to total protein concentration. The maximum value of v is N, the number of binding sites.

$$v = \frac{[PL]}{[P]_{tot}} = \frac{NKL_f}{1 + KL_f} \tag{5.4}$$

The most common way to measure binding constants and number of sites is through a Scatchard plot of v/L_f vs. v, since the above equation can be rearranged to

$$\frac{v}{L_f} = NK_d - vK_d \tag{5.5}$$

The intercept on the v-axis (where $v/L_f = 0$) is N; the slope is $-K$; and the intercept on the v/L_f axis (where $v = 0$) is NK. Let us repeat the preceding calculation with a wider range of added ligand concentrations, and then assume that there is a 5% random error, normally distributed, in the measurements of L_f and v.

```
> bind1 = function(x, Pt, Lt, K) {
Lf = x[1]; Pf = x[2]; PL = x[3];
(Pt-Pf-PL)^2 + (Lt-Lf-PL)^2 + (Pf*Lf-K*PL)^2
}
> Ptot = 1; Ltot = c(.2,.5,1,2,5,10,20,50); Kd = 4
> y = function(i) optim(c(1, .5, .5), bind1,
  method = "L-BFGS-B", lower = c(0,0,0), upper =
  c(Ltot[i], Ptot, Ptot), Lt=Ltot[i], Pt=Ptot, K=Kd)
> ypar = function(i) y(i)$par
> yp = matrix(nrow=length(Ltot), ncol=2+length(Kd),
  byrow=T)
> for (i in 1:length(Ltot)) yp[i,] = y(i)$par
> yp
          [,1]      [,2]      [,3]
[1,]   0.1612  0.96125  0.03875
[2,]   0.4075  0.90754  0.09246
[3,]   0.8284  0.82843  0.17157
[4,]   1.7016  0.70156  0.29844
[5,]   4.4721  0.47214  0.52786
[6,]   9.3007  0.30074  0.69926
```

```
[7,] 19.1726 0.17262 0.82738
[8,] 49.0754 0.07536 0.92464
```

The concentration of free ligand, `Lf`, is in the first column, and the concentration of bound ligand, `PL`, is in the third column. We apply random errors to `Lf` and `v` with `rnorm` of mean 0 and standard deviation 5%, or .05.

```
> Lf = yp[,1]; Pf = yp[,2]; PL = yp[,3]; v = PL/(Pf+PL)
> vexper = v*(1+rnorm(length(Ltot),0,.05))
> Lfexper = Lf*(1+rnorm(length(Ltot),0,.05))
```

A linear least-squares fit to the Scatchard plot is obtained by the `lm` (linear model) function:

```
> lm(vexper/Lfexper~vexper)

Call:
lm(formula = vexper/Lfexper ~ vexper)

Coefficients:
(Intercept)          vexper
      0.250          -0.255
```

A Scatchard plot of the "experimental" data, with the straight line corresponding to the computed slope and intercept, is then obtained with

```
> plot(vexper,vexper/Lfexper)
> abline(0.250, -0.255)
```

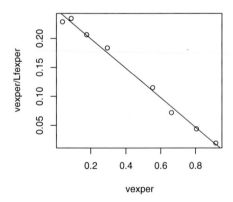

Fig. 5.1 Scatchard plot

The intercept on the x-axis is approximately 1, as it should be. The dissociation constant K is 1/.255 from the slope, close (but not identical, due to "experimental" uncertainties) to the input value of 4. Of course, another simulation would give slightly different results, because of the different random numbers that would be generated.

We can integrate the linear fitting and drawing the best-fit line as follows:

```
> Scatchard.lm = lm(vexper/Lfexper~vexper)
> lines(vexper,predict(Scatchard.lm))
```

This gives the same result as `abline`.

5.1.3 Strong and weak binding

Often, a protein will have several identical primary (strong) binding sites, and also some weaker ones. Here, for example, is the code (`bindsw` stands for binding-strong-weak) to calculate ligand binding for a protein with two strong sites ($K_d = 1$ and 4) and two weaker ones ($K_d = 50$ and 100). Note that in the code for y we have used the `rep` function, rather than repeating 0 or `Ptot` 5 or 6 times. We have also made the initial guesses 1/2 of the starting amounts of P and L.

```
> bindsw = function(x, Pt, Lt, K) {
Lf=x[1]; Pf=x[2]; PL=x[3]; PL2=x[4]; PL3=x[5]; PL4=x[6]
(Pt-Pf-PL-PL2-PL3-PL4)^2 + (Lt-Lf-PL-2*PL2-3*PL3-4*PL4)^2 +
(Pf*Lf-K[1]*PL)^2 + (PL*Lf-K[2]*PL2)^2 + (PL2*Lf-K[3]*PL3)^2 +
(PL3*Lf-K[4]*PL4)^2
}

> Ptot=1; Ltot=c(.5,1,2,5,10,20, 50, 100); Kd=c(1,4,50,100)
> y = function(i) optim(0.5*c(Ltot[i], rep(Ptot,5)), bindsw,
  method = "L-BFGS-B", lower = rep(0,6), upper = c(Ltot[i],
  rep(Ptot,5)), Pt=Ptot, Lt=Ltot[i], K=Kd)
> ypar = function(i) y(i)$par
> yp = matrix(nrow=length(Ltot), ncol=2+length(Kd), byrow=T)
> for (i in 1:length(Ltot)) yp[i,] = y(i)$par
> v = (yp[,3]+2*yp[,4]+3*yp[,5]+4*yp[,6])/Ptot
> v
[1] 0.2344 0.4372 0.7697 1.3460 1.7609 2.1421 2.7058 3.1631
> Lf = yp[,1]
> Lf
[1]   0.2656   0.5609   1.2303   3.6540   8.2390 17.8579 47.2918
   96.8369
> plot(v,v/Lf); abline(1,-0.5)
```

The Scatchard plot shows characteristic upward curvature and a rather inconclusive tail at high v, indicative of heterogeneous binding sites. Extrapolating the first few points—those dominated by the initial strong binding—to the v-axis gives an intercept of 2, consistent with two strong binding sites. The slope is -0.5, corre-

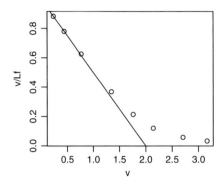

Fig. 5.2 Scatchard plot showing strong and weak binding

sponding to the "intrinsic" dissociation constant 0.5 of a single site; and the intercept
on the y-axis is 1 ($NK = 2 \times 0.5$).

5.1.4 Cooperative binding

Some proteins exhibit cooperative binding, in which the binding of the first ligand
enhances the binding of subsequent ligands. The binding of four molecules of oxygen by the hemoglobin tetramer is the best-known example. We consider here a
simpler example with just two sites.

Before embarking on a specific numerical example, we note that if P has two
identical binding sites for ligands, then we would expect that the two dissociation
constants are identical. However, a proper treatment takes into account "statistical
factors". There is only one empty site at which to bind the second ligand, compared to two for the first ligand; and there are two ways to dissociate a ligand from
PL2, and only one way from PL. Therefore, the dissociation constant for the second
binding event will be four times that for the first. In the following example, the first
dissociation constant is 4 (in appropriate concentration units) and the second is 1.
Thus the second binding site is actually 16 times as strong as the first.

```
> bind2 = function(x, Pt, Lt, K) {
Lf = x[1]; Pf = x[2]; PL = x[3]; PL2 = x[4];
(Pt-Pf-PL-PL2)^2 + (Lt-Lf-PL-2*PL2)^2 +
(Pf*Lf-K[1]*PL)^2 + (PL*Lf-K[2]*PL2)^2}

> Ptot = 1; Ltot = c(5,10,20, 50, 100); Kd = c(4,1)
```

```
> y = function(i) optim(0.5*c(Ltot[i], Ptot, Ptot,
  Ptot), bind2, method = "L-BFGS-B",
  lower = c(0,0,0,0), upper = c(Ltot[i], Ptot, Ptot,
  Ptot), Pt=Ptot, Lt=Ltot[i], K=Kd)
> ypar = function(i) y(i)$par
> yp = matrix(nrow=length(Ltot), ncol=2+length(Kd),
  byrow=T)
> for (i in 1:length(Ltot)) yp[i,] = y(i)$par
> yp
         [,1]       [,2]      [,3]     [,4]
[1,]    3.570 0.196933 0.17574 0.6273
[2,]    8.204 0.050311 0.10318 0.8465
[3,]   18.075 0.011469 0.05182 0.9367
[4,]   48.024 0.001696 0.02036 0.9779
[5,]   98.013 0.000000 0.01007 0.9883

> v = (yp[,3]+2*yp[,4])/Ptot
> v
[1] 1.430 1.796 1.925 1.976 1.987

> plot(Ltot,v,type="o")
```

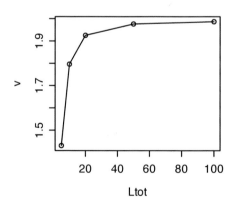

Fig. 5.3 Binding in a cooperative two-site system

Next we do a Scatchard plot, with an expanded range of Ltot:

```
> Ltot = c(.5,1,2,5,10,20, 50, 100)
> Lf = yp[,1]
```

```
> Lf
[1]    3.570   8.204 18.075 48.024 98.013
> plot(v,v/Lf, type="o", main="Scatchard Plot")
```

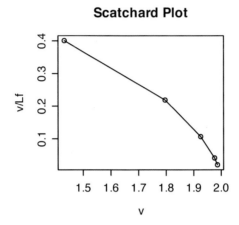

Fig. 5.4 Scatchard plot for a cooperative two-site system

The curvature of the Scatchard plot, concave to the x-axis, is characteristic of cooperative binding.

5.1.5 Oxygen binding by hemoglobin

As noted above, binding of oxygen by hemoglobin is the paradigmatic and best-studied case of cooperative ligand binding. Hemoglobin has four subunits, each of which can bind one O_2. There are five species, unliganded Hb and Hb with 1–4 O_2s bound, whose concentrations are related to each other through four equilibrium expressions:

$$K_i = \frac{pO_2[HbO_{2,i-1}]}{[HbO_{2,i}]}, \quad i = 1\ldots N \tag{5.6}$$

where the concentration of oxygen is conventionally measured as the partial pressure of the gas, pO_2, $N = 4$ for the hemoglobin system, and K_i is the dissociation constant for the ith step.

K includes the statistical factors. The relation to the intrinsic dissociation constant k is

$$K_i = \frac{i}{N-i+1}k_i \tag{5.7}$$

The average extent of oxygen binding is given as the quantity

$$Y = \frac{v}{N} = \frac{1}{4} \frac{[HbO_2] + 2[Hb(O_2)_2] + 3[Hb(O_2)_3] + 4[Hb(O_2)_4]}{[Hb] + [HbO_2] + [Hb(O_2)_2] + [Hb(O_2)_3] + [Hb(O_2)_4]} \qquad (5.8)$$

This is the number of ligands bound per Hb tetramer. Substituting eq. 5.6 and simplifying, we obtain

$$Y = \frac{1}{4} \frac{pO_2/k_1 + 2pO_2^2/k_1k_2 + 3pO_2^3/k_1k_2k_3 + 4pO_2^4/k_1k_2k_3k_4}{1 + pO_2/k_1 + pO_2^2/k_1k_2 + pO_2^3/k_1k_2k_3 + pO_2^4/k_1k_2k_3k_4} \qquad (5.9)$$

These equations and values of k determined by I. Tyuma, K. Imai, and K. Shimizu, *Biochemistry* 12:1491 (1973)

```
> k.noSalt = c(88,61,8.5,2.5) # 10 mM Tris,pH 7.4,25C
> k.Salt = c(420, 130, 120, 1.4) # + 0.1 M NaCl
```

enable us to calculate the binding curves using R. First we define the function that converts between k and K:

```
> K = function(k) {
K = c() # Initialize the vector
N = length(k) # How many values?
# Statistical factor
for (i in 1:N) K[i] = i/(N-i+1)*k[i]
return(K) # Get complete vector
}
```

Then we write a function to calculate Y at a given L from eq. 5.9. (To simplify typing, we substitute L (ligand) for pO_2.)

```
> Yi = function(L,K) {
N = length(K)
conc = c()
conc[1] = L/K[1]
for (i in 2:N) conc[i] = conc[i-1]*L/K[i]
numer = sum((1:N)*conc)/N
denom = 1 + sum(conc)
return(numer/denom)
}
```

Then we write a function to calculate the values of Y corresponding to the range of L values.

```
> Y = function(L,K) {
YY= c()
for (j in 1:length(L)) YY[j] = Yi(L[j],K)
return(YY)
}
```

Finally, we calculate and plot curves of Y vs. L for the no-salt and salt cases.

```
> L = seq(1,201,2)

> Y.noSalt = Y(L,K(k.noSalt))
> Y.Salt = Y(L, K(k.Salt))
> plot(L,Y.noSalt,type="l",xlab = "pO2", ylab = "Y")
> lines(L,Y.Salt, lty=2)
> legend(100,0.7,legend=c("No salt","0.1 M NaCl"),
  lty=c(1,2), bty="n")
```

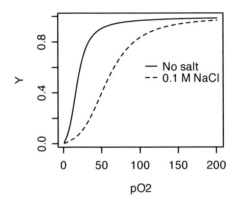

Fig. 5.5 Binding of O_2 to hemoglobin in the presence and absence of salt

The plot exhibits the sigmoidal shape characteristic of cooperative binding. Even more dramatically indicative of cooperative behavior is the Scatchard-type plot

```
> plot(Y.noSalt, Y.noSalt/L, type="l", xlab="Y",
  ylab="Y/pO2")
```

5.1.6 Hill plot

The ultimate in cooperative binding would occur if the only protein species were the unliganded and fully liganded forms. This is not known to occur in nature, but this simplified model—known as the Hill model—is useful for some purposes. We consider the reaction

$$P+NL \rightleftharpoons PL_N \tag{5.10}$$

with equilibrium dissociation constant (units of concentration N)

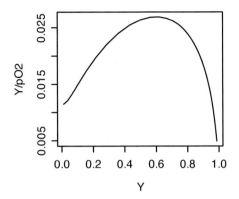

Fig. 5.6 Scatchard plot for binding of O_2 to hemoglobin in the absence of salt

$$K_d = \frac{[P][L]^N}{[PL_N]}.$$

(5.11)

The fractional saturation $Y = v/N$ is

$$Y = \frac{1}{N} \frac{N[PL_N]}{[P] + [PL_N]}$$

(5.12)

which in light of eq. 5.11 can be written

$$Y = \frac{[L]^N/K_d}{1 + [L]^N/K_d}$$

(5.13)

Thus the ratio of fraction liganded to fraction unliganded can be written

$$\frac{Y}{1-Y} = [L]^N/K_d$$

(5.14)

and a plot of $\ln[Y/(1-Y)]$ vs. $\ln[L]$ has a slope of N.

In any real binding situation, the slope will be equal to 1 at the extremes, and rise to a maximum near the midpoint of the binding isotherm. With hemoglobin as an example, we first generate the Hill plot and then numerically differentiate to obtain the slope.

```
> par(mfrow=c(1,2)) # Plot two graphs side by side
> plot(log10(L),log10(Y.noSalt/(1-Y.noSalt)), type="l",
  xlab="log(pO2)", ylab="log(Y/(1-Y))") # Hill plot
# Numerical differentiation
> dx = log10(L)[2:length(L)]-log10(L)[1:(length(L)-1)]
```

```
> dy = log10(Y.noSalt/(1-Y.noSalt))[2:length(L)]-
log10(Y.noSalt/(1-Y.noSalt))[1:(length(L)-1)]
# Slope of Hill plot
> plot(log10(L)[1:(length(L)-1)], dy/dx, type="l",
xlab="log(pO2)", ylab="Slope")
> par(mfrow=c(1,1)) # Return to single graph mode
```

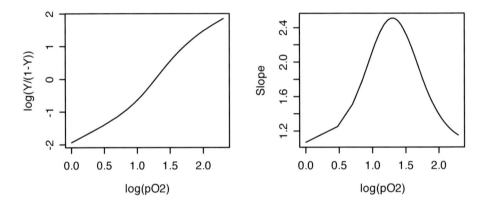

Fig. 5.7 (left) Hill plot for binding of O_2 to hemoglobin. (right) Dependence of the Hill plot slope on $\log(pO_2)$

Note that even for this very cooperative system, the Hill coefficient never approaches a maximum of 4, but instead peaks a bit above 2.4.

5.1.7 Experimental determination of equilibrium constants

We turn now to a brief discussion of how to measure equilibrium constants. For simplicity, we consider the reaction $P \rightleftharpoons Q$. Typically we measure some property of the system, e.g., optical density or fluorescence intensity, to which both P and Q contribute but at different levels. For concreteness, we will use the optical absorbance A at a particular wavelength. The contributions of pure P and pure Q, which can be measured separately, are A_P and A_Q. We also assume that that A is linear in the concentrations of P and Q.

Then, if f is the equilibrium fraction of material in the Q form, and $1 - f$ the fraction in the P form, the absorbance of the equilibrium mixture is

$$A = (1 - f)A_P + fA_Q \qquad (5.15)$$

This equation can be rearranged to solve for f:

$$f = \frac{A - A_P}{A_Q - A_P} \tag{5.16}$$

The equilibrium constant for the reaction is

$$K = \frac{[B]}{[A]} = \frac{f}{1 - f} \tag{5.17}$$

Combining eqs. 5.16 and 5.17 we obtain

$$K = \frac{A - A_P}{A_Q - A} \tag{5.18}$$

A might be measured at several wavelengths (at which A_P and A_Q, and thus A, differ) to provide data for a more statistically reliable estimate of K.

5.1.8 Temperature dependence of equilibrium constants: the van't Hoff equation

The equilibrium constant changes with temperature according to the van't Hoff equation

$$K(T) = K(T_0) \exp\left\{-\frac{\Delta H^0}{R}\left[\frac{1}{T} - \frac{1}{T_0}\right]\right\} \tag{5.19}$$

where ΔH^0 is the standard enthalpy of the reaction, R is the gas constant (8.314 J/mol), and T_0 is a reference temperature. If $\Delta H^0 < 0$, the reaction is *exothermic* (gives off heat), and K will decrease as T increases. If $\Delta H^0 > 0$, the reaction is *endothermic*, and K will increase as T increases. Many biochemical reactions are sensitive to temperature, which is why temperatures outside the normal range, such as a fever, can be so dangerous.

According to thermodynamics, the standard Gibbs energy ΔG^0 is related to the equilibrium constant by

$$\Delta G^0 = -RT \ln K \tag{5.20}$$

and to the standard enthalpy ΔH^0 and entropy ΔS^0 by

$$\Delta G^0 = \Delta H^0 - T\Delta S^0 \tag{5.21}$$

Thus if K and its temperature dependence can be measured, the thermodynamic functions ΔG^0, ΔH^0, and ΔS^0 can all be calculated.

5.2 Single strand–double helix equilibrium in oligonucleotides

A reaction of central importance in biochemistry and molecular biology is the association of single nucleic acid strands to form a double helix. This process leads to the DNA double helix, to DNA-RNA recognition during transcription, and to helical tracts in RNA. It also forms the basis for hybridization in Southern and Northern blotting and to the annealing of probes with targets in microarray analysis (see Chapter 14). Accordingly, in this section we show how to analyze the equilibrium between single-stranded and double-helical forms of DNA and RNA.

We begin with a review of some thermodynamic principles. The equilibrium is governed by the Gibbs free energy change ΔG, which is related to the chemical potentials μ of reactants and products. If we denote the single strands by X and Y, and the double helix by XY,

$$\Delta G = \mu_{XY} - \mu_X - \mu_Y \tag{5.22}$$

The chemical potentials are related to molar concentrations C by

$$\mu = \mu^0 + RT \ln C \tag{5.23}$$

where μ^0 is the standard chemical potential, R is the gas constant (1.987 cal/mol-K or 8.317 joule/mol-K), and T is the Kelvin temperature. Combining these equations gives

$$\Delta G = \mu_{XY}^0 - \mu_X^0 - \mu_Y^0 + RT \ln C_{XY} - RT \ln C_X - RT \ln C_Y \tag{5.24}$$

$$= \Delta G^0 + RT \ln \frac{C_{XY}}{C_X C_Y} \tag{5.25}$$

where ΔG^0, the standard Gibbs free energy change, is related to the standard enthalpy and entropy changes ΔH^0 and ΔS^0 by

$$\Delta G^0 = \Delta H^0 - T \Delta S^0 \tag{5.26}$$

At thermodynamic equilibrium, $\Delta G = 0$. The temperature at that point is the "melting temperature" T_m, so we find

$$\Delta H^0 - T_m \Delta S^0 = RT_m \ln \frac{C_X^{eq} C_Y^{eq}}{C_{XY}^{eq}} \tag{5.27}$$

so

$$T_m = \frac{\Delta H^0}{\Delta S^0 + R \ln \frac{C_X^{eq} C_Y^{eq}}{C_{XY}^{eq}}} \tag{5.28}$$

We now assume that X and Y are present in equal amounts. At equilibrium at the midpoint of the transition, half of the strands are in the double helix, and the other half remain single-stranded. Since the double helix contains two strands, its *molar* concentration is $C_{XY}^{eq} = C_T/4$, where C_T is the total concentration of strands. If the single strands X and Y are different and not self-complementary, the equilibrium concentration of each is $C_T/4$. Thus we have

$$T_m = \frac{\Delta H^0}{\Delta S^0 + R\ln(C_T/4)}, \quad \text{X,Y not self-complementary} \tag{5.29}$$

On the other hand, if X and Y are identical and self-complementary (e.g., AAAUUU), the concentration of single-strand species is $C_X^{eq} = C_Y^{eq} = C_T/2$, so $C_X^{eq}C_Y^{eq} = C_T^2/4$ and

$$T_m = \frac{\Delta H^0}{\Delta S^0 + R\ln(C_T)}, \quad \text{X,Y self-complementary} \tag{5.30}$$

We are now in a position to calculate estimates of T_m for oligonucleotides of specified sequence, given the enthalpy and entropy of the single-strand to double-helix reaction. These can be calculated by summing up tabulated values for the individual base-pairing reactions as the double helix zips up, then adding corrections for initiation, termination, and self-complementarity.

The values as determined by SantaLucia et al. [53] for DNA oligonucleotides and Turner et al. [69] for RNA, and tabulated in [8, Ch. 8] are expressed in R vector form on p. 100.

dinuc.DNA gives the dinucleotide sequence, from 5' to 3', which is to be paired with its complement. For example, 5'-GC-3' base-pairs with 3'-CG-5'. Note that although there are 16 dinucleotides, only 10 are listed. The others are formed from the complements of those listed: CC is equivalent to GG, TG to CA, etc. These transformations are embodied in the code below.

The thermodynamic values in dH.DNA, dS.DNA, and dG.DNA are the standard enthalpy, entropy, and Gibbs free energy (at 37°C) for the base-pairing reaction of the second base in each dinucleotide. For example, ΔH^0 for forming the C:G base pair in the GC:CG duplex, given that the G:C pair is already formed, is -9.8 kcal/mol, and ΔS^0 is -24.4 cal/mol-K. Note that since ΔH^0 is in kcal, while ΔS^0 is in cal, we must multiply ΔH^0 by 1000 to obtain consistent units.

To form the first base pair in the double helix, there are initiation enthalpy and entropy of 0.2 kcal/mol and -5.6 cal/mol-K. If the last base-pair formed is A:T there is a termination enthalpy of 2.2 kcal/mol and entropy of 6.9 cal/mol-K. These initiation and termination "penalties" reflect the difficulty of starting the double helix (particularly the entropic cost) and the weaker base stacking of A:T relative to G:C. There are additional penalties if there are sequence mismatches leading to internal loops or bulges, but we shall not consider them here.

Our goal is to write a function tm that calculates the melting temperature, given the oligonucleotide sequence, whether it is DNA or RNA, and the concentration of strands. Before writing that function, we write two others: dnvec to transform the sequence to a vector of dinucleotides, and selfcomp to check the sequence for self-complementarity. These functions will be called from the main function tm.

```
# Function to split sequence NAseq and transform to
# vector of dinucleotides dn
```

```
# DeltaH0 (kcal), DeltaS0 (eu), and DeltaG0.37 (kcal) values for DNA
dinuc.DNA = c("GC", "GG", "CG", "GA", "GT", "CA", "CT", "TA", "AT", "AA")
dH.DNA = c(-9.8,  -8.0,  -10.6,  -8.2,  -8.4,  -8.5,  -7.8,  -7.2,  -7.2,  -7.9)
dS.DNA = c(-24.4, -19.9, -27.2, -22.2, -22.4, -22.7, -21.0, -21.3, -20.4, -22.2)
dG.DNA = c(-2.24, -1.84, -2.17, -1.30, -1.44, -1.45, -1.28, -0.58, -0.88, -1.00)
init.DNA = c(0.2,  -5.6,  1.96)  # dH0, dS0, dG0.37
term.DNA = c(2.2,  6.9,  0.05)   # dH0, dS0, dG0.37 # For terminal AT pair


# DeltaH0 (kcal), DeltaS0 (eu), and DeltaG0.37 (kcal) values for RNA
dinuc.RNA = c("GC", "GG", "CG", "GA", "GU", "CA", "CU", "UA", "AU", "AA")
dH.RNA = c(-14.88,-13.39,-10.64,-12.44,-11.40,-10.48,-10.44,-7.69,-9.38,-6.82)
dS.RNA = c(-36.9, -32.7, -26.7, -32.5, -29.5, -27.1, -26.9, -20.5, -26.7, -19.0)
dG.RNA = c(-3.42, -3.26, -2.36, -2.35, -2.24, -2.11, -2.08, -1.33, -1.10, -0.93)
init.RNA = c(3.61, -1.5,  4.09)  # dH0, dS0, dG0.37
term.RNA = c(3.72, 10.5,  0.45)  # dH0, dS0, dG0.37 # For terminal AU pair
```

```
dnvec = function(NAseq, DR) {
NAseq.split = strsplit(NAseq, split="")
NAseq = unlist(NAseq.split)
lseq = length(NAseq)
dn = c() # Dinucleotide vector
for (i in 1:(lseq-1)) {
dn[i] = paste(NAseq[i], NAseq[i+1], sep="")
if (DR == "d") {
# Substitute bottom strand for top
if (dn[i]=="CC") dn[i] = "GG"
if (dn[i]=="TC") dn[i] = "GA"
if (dn[i]=="AC") dn[i] = "GT"
if (dn[i]=="TG") dn[i] = "CA"
if (dn[i]=="AG") dn[i] = "CT"
if (dn[i]=="TT") dn[i] = "AA"
}
else {
# Substitute bottom strand for top
if (dn[i]=="CC") dn[i] = "GG"
if (dn[i]=="UC") dn[i] = "GA"
if (dn[i]=="AC") dn[i] = "GU"
if (dn[i]=="UG") dn[i] = "CA"
if (dn[i]=="AG") dn[i] = "CU"
if (dn[i]=="UU") dn[i] = "AA"
}
}
return(list(NAseq,dn))
}

# Function to check for self-complementarity
selfcomp = function(Nseq, DR) {
compseq = c()
if (DR == "d") {
for (i in 1:length(Nseq)){
if (Nseq[i] == "A") compseq[i] = "T" else
if (Nseq[i] == "T") compseq[i] = "A" else
if (Nseq[i] == "C") compseq[i] = "G" else
if (Nseq[i] == "G") compseq[i] = "C"
}
     }
else {
for (i in 1:length(Nseq)){
if (Nseq[i] == "A") compseq[i] = "U" else
if (Nseq[i] == "U") compseq[i] = "A" else
if (Nseq[i] == "C") compseq[i] = "G" else
if (Nseq[i] == "G") compseq[i] = "C"
```

```
}
      }
# selfcomp = 1 if the sequence is self-complementary
# That is, the reverse of the complementary sequence
  = the original sequence
# Otherwise, selfcomp = 0
selfcomp = sum(Nseq == rev(compseq)) == length(Nseq)
}

# Main function to calculate tm for a sequence, given
# sequence, total conc, and RNA/DNA
tm = function(NAseq, CT, DR) {
# Check whether "r" or "d" is specified
if (DR !="r" & DR != "d") stop("Must specify r or d")

# Call function dnvec to split sequence and transform
# to dinucleotides
Nseq = dnvec(NAseq,DR)[[1]]; lseq = length(Nseq)
dn = dnvec(NAseq, DR)[[2]]; ldn = length(dn)

# Choose thermo functions appropriate for RNA or DNA
if (DR == "r") {dinuc=dinuc.RNA; dH=dH.RNA; dS=dS.RNA;
init = init.RNA; term = term.RNA}
else {dinuc=dinuc.DNA; dH=dH.DNA; dS=dS.DNA;
init = init.DNA; term = term.DNA}

# Calculate dH, dS from sequence before corrections
# for init, AT, self-comp
dH0 = c(); dS0 = c()
for (i in 1:ldn) {
  dni = dn[i]
  dH0[i] = dH[dinuc == dni]
  dS0[i] = dS[dinuc == dni]
  }
dH0.tot = sum(dH0)
dS0.tot = sum(dS0)

# Add initiation parameters
dH0.tot = dH0.tot + init[1]
dS0.tot = dS0.tot + init[2]

# Check for terminal AU or AT
# ATU = 1 if terminal residue is A, T, or U;
# 0 otherwise
ATU = Nseq[lseq] == "A" | Nseq[lseq] == "T" |
  Nseq[lseq] == "U"
```

```
dH0.tot = dH0.tot + ATU*term[1]
dS0.tot = dS0.tot + ATU*term[2]

# Call function to check for self-complementarity.
# If so, decrease entropy by Rln(2) = 1.4 cal/mol-K
# to account for symmetry
# 1 if self-complementary, 0 otherwise
sc = selfcomp(Nseq, DR)
dS0.tot = dS0.tot -1.4*sc

# Return tm, given  total strand concentration CT
# Assumes the two strands are in equal concentrations
R = 1.987 # cal/mol-deg Gas Constant
if (sc == 0) CT = CT/4 # Effective conc if not self-comp

tm = 1000*dH0.tot/(dS0.tot + R*log(CT)) - 273.15 # DegC
return(tm)
}

# Example
NAseq = "CCGG"; CT = 1e-4; DR = "r"
tm(NAseq, CT, DR)
[1] 25.25881 # Compared with 27.1 experimental
```

5.3 Steady-state enzyme kinetics

5.3.1 Michaelis-Menten kinetics

The equation for the velocity of an enzyme-catalyzed reaction that obeys Michaelis-Menten kinetics is of the same hyperbolic form as the equation for the extent of binding of a ligand to a protein. This is not surprising since they are both saturable binding processes.

$$v = \frac{V_{max}[S]}{K_m + [S]} \text{ compared with } v = \frac{NKL_f}{1 + KL_f} \tag{5.31}$$

The conventions for plotting enzyme kinetic and binding data are somewhat different, however. Instead of the Scatchard equation, it is common to use either the double-reciprocal Lineweaver-Burk plot of $1/v$ vs. $1/[S]$

$$\frac{1}{v} = \frac{1}{V_{max}} + \frac{K_m}{V_{max}} \frac{1}{[S]} \tag{5.32}$$

or the Eadie-Hofstee plot of v vs. $v/[S]$, which is just the Scatchard equation turned on its side

$$v = V_{max} - K_m \frac{v}{[S]}$$ (5.33)

We will simulate enzyme kinetic data, with realistic errors, and plot the data in both forms. We let $V_{max} = 10$, $K_m = 10$, and $[S]$ run from 1 to 100. As with ligand binding, the units are arbitrary but consistent. We assume 5% error of measurement in both v and $[S]$, and plot error bars as described in Chapter 2 on graphing.

```
> Vmax = 10; Km = 10;
> S = c(1,2,5,10,20,50,100)*(1+rnorm(7,0,.05))
> v = Vmax*S/(Km + S)*(1+rnorm(7,0,.05))
> plot(S,v,type="b",xlim=c(0,100),ylim=c(0,10))
> arrows(S,v,S,v+.05*v,.05,90)
> arrows(S,v,S,v-.05*v,.05,90)
> arrows(S,v,S+.05*S,v,.05,90)
> arrows(S,v,S-.05*S,v,.05,90)
```

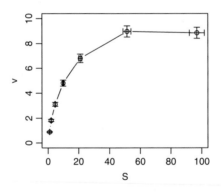

Fig. 5.8 Velocity vs. substrate concentration for simulated enzyme kinetic data

5.3.2 Lineweaver-Burk fitting

The Lineweaver-Burk analysis goes as follows:

```
> vrecip = 1/v; Srecip = 1/S
> LwB = lm(vrecip ~ Srecip) # Linear least squares fit
```

```
> LwB

Call:
lm(formula = vrecip ~ Srecip)

Coefficients:
(Intercept)              Srecip
     0.0867              1.0937

> plot(Srecip,vrecip,main="Lineweaver-Burk Plot",
  xlab="1/v", ylab="1/S")
> abline(LwB)
```

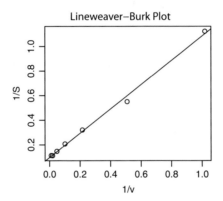

Fig. 5.9 Lineweaver-Burk plot for simulated enzyme kinetic data

The line representing the best fit slope and intercept is drawn with the abline function using the parameters provided by LwB. A statistical analysis of the fit is obtained using summary(LwB):

```
> summary(LwB)

Call:
lm(formula = 1/v ~ Srecip)

Residuals:
          1          2          3          4          5
   0.016520  -0.029600  -0.020128   0.011242   0.000224
          6          7
```

```
 0.011070   0.010672
```

```
Coefficients:
             Estimate Std. Error t value Pr(>|t|)
(Intercept)   0.08673    0.00943     9.2  0.00025 ***
Srecip        1.09366    0.02211    49.5  6.4e-08 ***
---
Signif. codes:  0 '***' 0.001 '**' 0.01 '*' 0.05 '.'
0.1 ' ' 1
```

```
Residual standard error: 0.0196 on 5 degrees of freedom
Multiple R-Squared: 0.998, Adjusted R-squared: 0.998
F-statistic: 2.45e+03 on 1 and 5 DF,  p-value: 6.38e-08
```

The standard errors in the estimates of slope and intercept are in the range 10%–20%. (Remember that S and v were uncertain by ±5%.) The fitted values of Vmax and Km are obtained from

```
> LwB.c = LwB$coefficients
> LwB.c
(Intercept)        Srecip
    0.08673       1.09366
> Vmax.app = 1/LwB.c[1]; Vmax.app
(Intercept)
       11.53
> Km.app = LwB.c[2]/LwB.c[1];  Km.app
Srecip
 12.61
```

Vmax.app and Km.app should be compared to the input values of 10 and 10.

5.3.3 Eadie-Hofstee fitting

The same process in R, with different dependent and independent variables, leads to the Eadie-Hofstee treatment.

```
> v.by.S = v/S
> EH = lm(v ~ v.by.S)
> EH
```

```
Call:
lm(formula = v ~ v.by.S)
```

```
Coefficients:
(Intercept)          v.by.S
```

```
        10.3              -10.6

> summary(EH)

Call:
lm(formula = v ~ v.by.S)

Residuals:
      1       2       3       4       5       6       7
 -0.621   0.161   0.615  -0.372   0.508  -0.191  -0.100

Coefficients:
            Estimate Std. Error t value Pr(>|t|)
(Intercept)    10.31       0.38    27.1  1.3e-06 ***
v.by.S        -10.60       0.67   -15.8  1.8e-05 ***
---
Signif. codes:  0 '***' 0.001 '**' 0.01 '*' 0.05 '.'
0.1 ' ' 1

Residual standard error: 0.497 on 5 degrees of freedom
Multiple R-Squared: 0.98, Adjusted R-squared: 0.976
F-statistic:  250 on 1 and 5 DF,  p-value: 1.84e-05

> plot(v.by.S, v, xlab="v/S", ylab="v",
  main="Eadie-Hofstee fit")
> abline(EH)
```

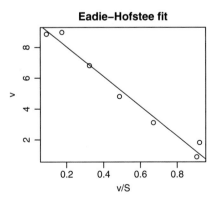

Fig. 5.10 Eadie-Hofstee plot for simulated enzyme kinetic data

The data look noisier; but we note that the apparent kinetic parameters, `Vmax.app` = 10.3 and `Km.app` = 10.6, are closer to the input values of 10 and 10 than they were in the Lineweaver-Burk plot, and the standard errors in the parameters are only about 4%–7%. This is because, as the plot shows, the experimental points are fairly evenly spread out. In the Lineweaver-Burk procedure, most of the points are bunched near the origin, while the point at lowest S and v is far away, giving it undue influence in the least-squares fitting due to a "lever" effect.

It is useful to plot the residuals, i.e., the deviations between observed and predicted values, in order to see whether they seem to be random or to obey some systematic pattern (which indicates inadequate fit).

```
> par(mfcol=c(1,2))
> plot(residuals(LwB), main="Lineweaver-Burk")
> abline(0,0)
> plot(residuals(EH), main = "Eadie-Hofstee")
> abline(0,0)
```

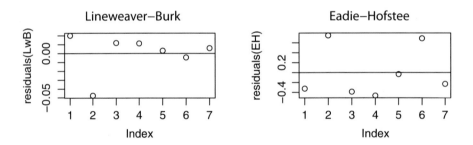

Fig. 5.11 Residuals for Lineweaver-Burk and Eadie-Hofstee fits

It will be left as a problem to add error bars to the Lineweaver-Burk and Eadie-Hofstee plots.

5.3.4 Competitive inhibition

To plot two sets of points on the same graph we use the `points` function. We will do this using the example of competitive inhibition, with both Lineweaver-Burk and Eadie-Hofstee representations. The equation for competitive inhibition is

$$v = \frac{V_{max}[S]}{K_m \left(1 + \frac{1}{K_i}\right) + [S]} \tag{5.34}$$

```
> Vmax = 10; Km = 10; S = c(1,2,5,10,20,50,100)
> Ki = 20; I = 20
> v.0 = Vmax*S/(Km + S) # Without inhibitor
> v.i = Vmax*S/(Km*(1 + I/Ki) + S) # With inhibitor
> plot(1/S, 1/v.0, type="b", ylim=c(0,1.2),
  main="Lineweaver-Burk")
> points(1/S,1/v.i,pch=19, type="b")
> plot(v.0/S,v.0,type="b", xlim=c(0,1), ylim=c(0,10),
  main="Eadie-Hofstee")
> points(v.i/S,v.i,type="b", pch=19)
```

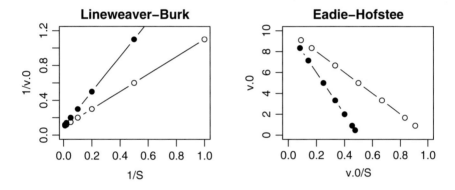

Fig. 5.12 Competitive inhibition plots according to Lineweaver-Burk and Eadie-Hofstee analysis

5.4 Non-linear least-squares fitting

Modern computers give us the power to avoid linearizing equations, if we so desire, and fit the parameters directly using non-linear least squares. This has the advantage, in principle, of weighting all the data equally. We illustrate with the enzyme kinetics data we have used throughout this section, using the nls function. (See R Help for details on how to use this function. Note the warning there not to use nls on artificial "zero-residual" data. Our kinetics data has random error that avoids this pitfall.)

```
> kin = nls(v ~ Vmax*S/(Km + S), start = list(Vmax = 1, Km = 1))
> summary(kin)
```

```
Formula: v ~ Vmax * S/(Km + S)

Parameters:
      Estimate Std. Error t value Pr(>|t|)
Vmax    10.047        0.251     40.1  1.8e-07 ***
Km       9.606        0.805     11.9  7.3e-05 ***
---
Signif. codes:  0 '***' 0.001 '**' 0.01 '*' 0.05 '.' 0.1 ' ' 1

Residual standard error: 0.235 on 5 degrees of freedom

> resid.kin = (v-predict(kin))
> resid.kin
[1] -0.1113  0.0197  0.0444 -0.2299  0.3751 -0.2500  0.0718
> plot(resid.kin); abline(0,0)
```

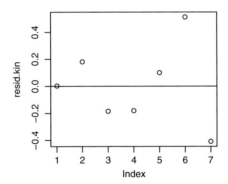

Fig. 5.13 Residuals from non-linear least-squares fitting to simulated enzyme kinetics data

Non-linear least-squares fitting is equally applicable to equilibrium binding data.

5.5 Problems

1. Verify the convergence difficulty stated in the first section/subsection when concentrations are given as micromolar or nanomolar. Modify the R code in this
   section to use the scaling method parscale for optim() described in R Help.
2. Verify the assertion, in the Successive binding of two ligands subsection, that a
   Scatchard plot will be properly linear if the second dissociation constant is 4×
   greater than the first.

3. Redo the Lineweaver-Burk and Eadie-Hofstee plots to include error bars and proper labeling and titling.

4. The pK_as of citric acid are 3.13, 4.76, and 6.40, according to the Merck Index. Calculate and plot a species distribution diagram for citric acid: the fractions of each species (H_3Cit, H_2Cit^-, $HCit^{-2}$, Cit^{-3}) for pH values between 1 and 9. Increment pH in units of 0.1 to give reasonable smooth curves. Use different colors for the different species, and add a legend and suitable labeling/titling.

5. The fumaric acid (F) to maleic acid (M) interconversion is a much-studied biochemical reaction. The apparent reaction equilibrium can be written

$$F_T \rightleftharpoons M_T, K_{app} = \frac{M_T}{F_T}$$

$$F_T = FH_2 + FH^- + F^{-2}; M_T = MH_2 + MH^- + M^{-2}$$

where we note that the total amounts of F and M are divided among diprotonated, monoprotonated, and unprotonated forms. The pK_as for fumaric acid are 3.02 and 4.40; for maleic acid they are 3.48 and 5.11. Write an R program to calculate and plot the apparent equilibrium constant, K_{app}, as a function of pH from 1 to 7, assuming that $K_{app} = 1.0$ at pH 3.75. Also calculate and plot the fraction of total maleic acid over the 1–7 pH range.

6. Spectroscopy is often used to monitor the concentrations of chemical species and their equilibrium. Suppose we are interested in the equilibrium A $\rightleftharpoons$ B with K = [B]/[A]. The UV spectrum of compound A has a Gaussian band shape with a maximum intensity of 1.0 at 260 nm and a band width (sd) of 8 nm. The spectrum of compound B also is Gaussian, with a maximum intensity of 0.7 at 265 nm, and sd also = 8 nm. Plot the individual spectra and the sum of spectra if [A] = [B]. If you want to use multi-wavelength spectral analysis to follow the equilibrium as the temperature is changed, what 5 wavelengths would be appropriate? (There is no unique answer to this question; just select reasonable values.) Which wavelength would you not want to use?

7. Suppose that the reaction mixture is 90% A at 10°C (283 K). Also suppose that the enthalpy of reaction ΔH^0 is 16 kJ/mol, and that the T-dependence of the reaction follows the van't Hoff equation

$$K(T_2) = K(T_1) \exp\left[-\frac{\Delta H^0}{R}\left(\frac{1}{T_2} - \frac{1}{T_1}\right)\right]$$

At what temperature will the reaction mixture be 90% B? Choose six temperatures (283, the T that you've just calculated, and four intermediate) and plot the resulting solution spectra. (Remember that the gas constant $R = 8.314$ J/mol.) Construct a table of absorbances vs. T at the 5 wavelengths you chose in the previous problem.

8. Now work backwards to calculate the concentrations of A and B and the equilibrium constants at each of the 6 temperatures, and ΔH^0 from the T-dependence of K. To make your simulation more realistic, assume that the absorbance values

can be measured only to 1% accuracy, so apply normal random noise with *mean* = 0 and *sd* = 0.01 to each of the absorbances in the table in Problem 7. How do your "experimental" determinations compare with the input information? How will you combine the results from the 5 wavelengths?

9. Take the spectrum resulting from equimolar amounts of A and B in Problem 7, with data plotted every 0.5 nm between 220 nm and 300 nm, and add 5% normally-distributed random noise. Use the R function `nls ()` to fit the spectrum to a sum of two Gaussians, when you make initial guesses for the individual spectral parameters that are at least 20% different from those you input.

10. Repeat Problem 9 with Lorentzian rather than Gaussian functions, but with the same maxima and widths. Do you observe any difference in the ability to fit to noisy data?

11. Calculate the melting temperature of the RNA oligonucleotide GCAAUACG if the total strand concentration is 2×10^{-4} M.

Chapter 6
Differential Equations and Reaction Kinetics

In this chapter we will show how to numerically solve the kinetic rate equations of the sort that describe biochemical metabolism, microbial growth, and similar biological phenomena. These equations are called differential equations, and generally describe the change of concentrations or numbers of organisms (the dependent variables) as a function of time (the independent variable). Since there is only one independent variable (time), these are "ordinary differential equations", or ODEs. If there were two or more independent variables (e.g., time and positional coordinates), we would have "partial differential equations", or PDEs. In this book we'll deal mainly with ODEs.

6.1 Analytically solvable models

For the most part, the differential equations that represent complex biological phenomena cannot be solved in closed form, and require numerical solutions. However, there are some useful models that are simple enough to be solved analytically. We will deal with some of them in this section, before describing numerical methods.

6.1.1 Exponential growth

If the rate of growth is proportional to the number of organisms N that exist at time t, the differential equation

$$\frac{dN}{dt} = kN \quad \text{with} \quad N = N_0 \quad \text{at} \quad t = 0 \tag{6.1}$$

has solution

$$N(t) = N_0 e^{kt} \tag{6.2}$$

V. Bloomfield, *Computer Simulation and Data Analysis in Molecular Biology and Biophysics,*
Biological and Medical Physics, Biomedical Engineering, DOI: 10.1007/978-1-4419-0083-8_6,
© Springer Science + Business Media, LLC 2009

To visualize this solution in R, we use the code

```
> N0 = 2; k = .1 # Per minute
> t = 0:50 # Minutes
> N = N0*exp(k*t)
> plot(t,N,type="l",xlab="t/min",main="Exponential Growth")
```

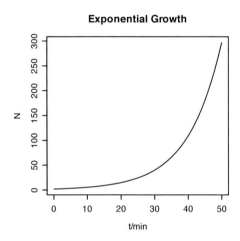

Fig. 6.1 Graph of exponential growth

6.1.2 *Exponential decay*

The opposite situation to exponential growth is exponential decay, in which the rate of decay is proportional to the population at time t. Exponential decay is often characterized by its half-life, the time to decay to one-half the initial population. The half-life $t_{1/2} = \ln(2)/k = 0.693/k$ where k is the decay rate, with units of reciprocal time. An example is radioactive decay. Consider the decay of ^{32}P, with $k = 0.04847$/day. The differential equation

$$\frac{dN}{dt} = -kN \quad \text{with} \quad N = N_0 \quad \text{at} \quad t = 0 \tag{6.3}$$

has solution

$$N(t) = N_0 e^{-kt} \tag{6.4}$$

```
> N0 = 500 # Microcuries
> k = 0.04847 # Per day
> t = 0:100
```

```
> N = N0*exp(-k*t)
> plot(t,N,type="l",xlab="t/day",main="Exponential Decay of P32")
> abline(h=N0/2,v=0.693/k,lty=3)
```

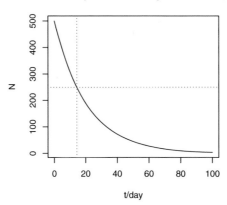

Fig. 6.2 Graph of exponential decay

6.1.3 Chemical Conversion

In a reaction A → B, A decays exponentially, and B grows exponentially. Suppose the concentration of A is initially 10 units, the concentration of B is 0, and the rate constant for conversion is 1/time.

```
> A0 = 10; k = 1; t = seq(0,5,by=.2)
> A = A0*exp(-k*t); B = A0*(1-exp(-k*t))
> plot(t,A,main="Conversion of A to B",type="l", ylab= "A, B")
> lines(t,B, lty=2)
> legend(3,6,legend=c("A","B"),lty=c(1,2),bty="n")
```

6.1.4 Exponential decay to a constant value

A number of simple models useful in various biological situations involve exponential decay to a final constant value rather than to zero. Examples are Newton's law of cooling from temperature T_0 to an external temperature T_f,

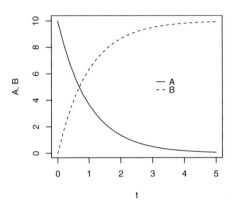

Fig. 6.3 Chemical conversions of A to B

$$\frac{dT}{dt} = -k(T - T_f) \quad \text{with} \quad T = T_0 \quad \text{at} \quad t = 0 \tag{6.5}$$

has solution

$$T(t) = T_f + (T_0 - T_f)e^{-kt} \tag{6.6}$$

and diffusion of a substance across a membrane from initial concentration C_0 into a reservoir with constant concentration C_f

$$\frac{dC}{dt} = -k(C - C_f) \quad \text{with} \quad C = C_0 \quad \text{at} \quad t = 0 \tag{6.7}$$

has solution

$$C(t) = C_f + (C_0 - C_f)e^{-kt} \tag{6.8}$$

Each of these has a time dependence of the form shown in Figure 6.4, obtained from the code

```
> k = 0.06; T0 = 37; Tf = 8
> t=0:48
> T = Tf+(T0-Tf)*exp(-k*t)
> plot(t,T,xlab="hours",ylab="body temp",
  main="Cooling Curve", type="l")
```

with parameters for the cooling of "an average clothed male corpse in still air" [39, p. 23].

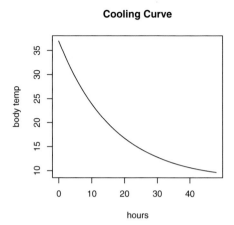

Fig. 6.4 Graph of exponential decay to a constant value

6.1.5 Model of limited growth

More realistic than unlimited exponential growth is a model in which some external factors (e.g., disease or limiting nutrient supply) constrain the maximum population to some size N_{max}. One simple model of this type can be written

$$\frac{dN}{dt} = kN(N_{max} - N) \quad \text{with} \quad N = N_0 \quad \text{at} \quad t = 0 \tag{6.9}$$

with solution

$$N(t) = \frac{N_0 N_{max}}{N_0 + (N_{max} - N_0)e^{-N_{max}kt}} \tag{6.10}$$

This model, implemented in the code

```
> N0 = 2; Nmax = 500; k = 4e-4 # Per individual per week
> t = 0:52 # Weeks
> N = N0*Nmax/(N0+(Nmax-N0)*exp(-Nmax*k*t))
> plot(t,N,xlab="weeks",main="Density-Limited Growth", type="l")
```

leads to the characteristic S-shaped growth curve shown in Figure 6.5.

6.1.6 Kinetics of bimolecular reactions

The reaction of two species A and B to form product P, A + B → P, is governed by the rate law

$$\frac{dB}{dt} = -k[A][B] = -k[B]^2 \text{ if } [A] = [B] \tag{6.11}$$

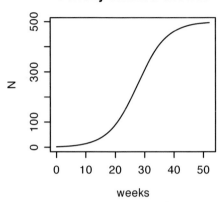

Fig. 6.5 Graph of density-limited growth

This integrates to

$$\frac{1}{[B]} - \frac{1}{[B]_0} = kt \tag{6.12}$$

which rearranges to

$$[B] = \frac{[B]_0}{1 + [B]_0 kt} \tag{6.13}$$

```
> k = 0.2; t = 0:20; B0 = 5 # Molar
> B = B0/(1+B0*k*t)
> plot(t,B, xlab="s",ylab="[B]/M",
  main="Bimolecular [A]=[B]",type="l")
> lines(t,1/B,lty=2)
> legend(4,4,legend=c("[B]","1/[B]"),lty=c(1,2),bty="n")
```

If $[A]_0 \gg [B]_0$, then $[A]$ hardly changes during the reaction, and $k[A]_0$ becomes a new pseudo-first-order rate constant k' so that $[B]$ decays exponentially. If $[A]_0$ is not equal to $[B]_0$ but is not much greater than $[B]_0$, the expression for the concentration changes is more complicated. Let x be the number of moles of A or B that has reacted, with $x = 0$ initially. The bimolecular rate law can then be written

$$\frac{dx}{dt} = k([A]_0 - x)([B]_0 - x) \tag{6.14}$$

which rearranges to

$$\frac{dx}{([A]_0 - x)([B]_0 - x)} = kdt \tag{6.15}$$

The result, after integration by parts of the left-hand side, is

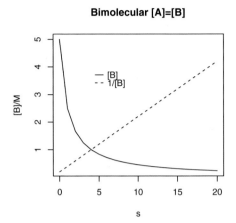

Fig. 6.6 Kinetics of a bimolecular reaction

$$\frac{1}{[A]_0 - [B]_0} \ln \frac{[B]_0([A]_0 - x)}{[A]_0([B]_0 - x)} = kt \tag{6.16}$$

6.2 Numerical integration of ODEs

If the differential equation(s) are too complicated to be solved analytically, numerical methods are required. The simplest—and often quite satisfactory—approach is Euler's method. There may be times when Euler's method does not give satisfactory results, in which case we'll describe some alternatives.

6.2.1 Integrating a single ODE by the Euler method

The basic idea of numerically integrating a differential equation of the form

$$\frac{dy}{dx} = f(x) \tag{6.17}$$

is to start at a point x_{i-1} where y is known to be y_{i-1}, take a small step Δx, and estimate the new value of $y_i = y_{i-1} + \Delta y$ where

$$\Delta y = \left(\frac{\Delta y}{\Delta x}\right) \Delta x \tag{6.18}$$

The Euler method implements this idea directly. We see an example in the following code for integrating $(dy/dx) = x^2$, which should yield $x^3/3$.

```
> # Integrating ODE by Euler method
> y = function(yinit,x){
> y = c(yinit); # Initial value passed to function
> # Divide x interval into n steps of length dx
> n = length(x); dx=(x[n]-x[1])/(n-1)
> for (i in 2:n) {
dy = f(x[i-1])*dx
y[i] = y[i-1]+dy
}
return(y)
}

> f = function(x) {x^2}

> # Divide x interval into n steps of length dx
> xinit = 0; xmax = 2; dx = .01; x = seq(xinit,xmax,dx)
> yinit = 0
> x1 = x
> y1 = y(yinit,x)
> plot(x1,y1, type="l")   # Plot the result as a line
```

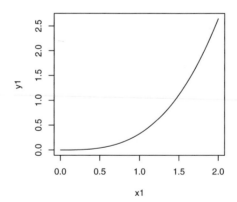

Fig. 6.7 Graph of exponential growth calculated by the Euler method

With x divided into 200 intervals of length 0.01, the Euler integration is very accurate, as you can verify by plotting the analytical value $x^3/3$ on the same graph.

However, if we used $dx = 0.1$, there would be a substantial deviation between numerical and analytical results (verify this). This points out the deficiency of the Euler method: it requires quite small increments to achieve high accuracy. This can be a problem for large problems, especially if the function is rapidly changing. However, the speed and 15-digit precision of modern personal computers makes this less of a problem than it used to be.

6.2.2 Integrating a system of ODEs by the Euler method

Next we show how to use the Euler method to solve a system of simultaneous differential equations, in this case the chemical interconversion $A \rightleftharpoons B \rightleftharpoons C$. We represent the concentrations as a function of time by using a matrix x, 3 rows (one for each species) by n columns (one for each time increment). We calculate the dx_i for all species first, then update all x_i to their new values. This is called the "two-step Euler method".

```
> # Forward and reverse rate constants
> kf1 = 1; kr1 = 3; kf2 = 5; kr2 = .1
> dt = .005; tmax = 10
> t = seq(0,tmax,dt); n = tmax/dt+1
> # All x(t) = 0 to start
> x = matrix(rep(0,3*n),nrow=3,ncol=n)
> x[1,1] = 1; x[2,1] = 0; x[3,1] = 0    # Initial values
> for(i in 2:n){
dx1 = -kf1*x[1,i-1]*dt+kr1*x[2,i-1]*dt;
dx2 = kf1*x[1,i-1]*dt-(kr1+kf2)*x[2,i-1]*dt+
        kr2*x[3,i-1]*dt;
dx3 = kf2*x[2,i-1]*dt-kr2*x[3,i-1]*dt;
x[1,i] = x[1,i-1]+dx1;
x[2,i] = x[2,i-1]+dx2;
x[3,i] = x[3,i-1]+dx3;
}
> plot(t,x[1,],type="l",lty=1,
  main="Interconversion of 3 species",
  xlab="t/s",ylab="Conc/M",ylim=c(0,1.025),bty="l")
> lines(t,x[2,],lty=2)
> lines(t,x[3,],lty=3)
> # Conservation of total species:
> lines(t,x[1,]+x[2,]+x[3,],lty=1)
> text(1,0.9,"x1")
> text(1,0.15,"x2")
> text(4,0.9,"x3")
> text(5,1.05,"Total")
```

Interconversion of 3 species

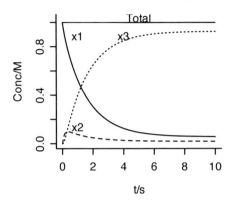

Fig. 6.8 Interconversion of three species calculated by the Euler method

6.2.3 Integrating a system of ODEs by the improved Euler method

Most of the error in the simple Euler method applied to rapidly changing functions occurs because it assumes that the slope dy/dt remains constant between t_{i-1} and t_i, while in fact the slope is a function of $y(t)$. The improved Euler method (also known as the second-order Runge-Kutta method) finds a better estimate of the slope by estimating dy/dt at t_{i-1}, using that estimate to find the slope at t_i, then averaging those two estimates. This procedure is utilized in the following code, which simulates the same $A \rightleftarrows B \rightleftarrows C$ reaction sequence.

```
> x1 = c(1); x2 = c(0); x3 = c(0)
> kf1 = 1; kr1 = 3; kf2 = 5; kr2 = .1
> dt = .05; tmax = 10
> t = seq(0,tmax,dt); n = tmax/dt+1
> for(i in 2:n){
x1a = x1[i-1]; x2a = x2[i-1]; x3a = x3[i-1];

# The first estimate of slopes

dx1a = -kf1*x1a*dt+kr1*x2a*dt;
dx2a = +kf1*x1a*dt-(kr1+kf2)*x2a*dt+kr2*x3a*dt;
dx3a = +kf2*x2a*dt-kr2*x3a*dt;

# The second estimate of slopes

x1b = x1a+dx1a; x2b = x2a+dx2a; x3b = x3a+dx3a;
```

```
dx1b = -kf1*x1b*dt+kr1*x2b*dt;
dx2b = +kf1*x1b*dt-(kr1+kf2)*x2b*dt+kr2*x3b*dt;
dx3b = +kf2*x2b*dt-kr2*x3b*dt;

# Averaging first and second estimate of slopes

dx1 = (dx1a+dx1b)/2
dx2 = (dx2a+dx2b)/2
dx3 = (dx3a+dx3b)/2

x1[i] = x1[i-1]+dx1
x2[i] = x2[i-1]+dx2
x3[i] = x3[i-1]+dx3
}
> plot(t,x1,type="l",lty=1,
  main="Interconversion of 3 species",
  xlab="t/s",ylab="Conc/M",ylim=c(0,1.025),bty="l")
> lines(t,x2,lty=2)
> lines(t,x3,lty=3)
> lines(t,x1+x2+x3,lty=1)
> text(1,0.9,"x1")
> text(1,0.15,"x2")
> text(4,0.9,"x3")
> text(5,1.05,"Total")
```

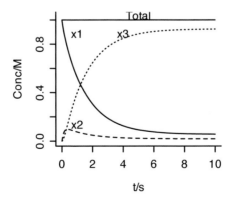

Interconversion of 3 species

Fig. 6.9 Interconversion of three species calculated by the improved Euler method

Note that the results are essentially the same, although the time step *dt* is 10 times as large in the improved Euler method as the simple Euler method.

6.2.4 Integrating a system of equations using 4th-order Runge-Kutta

Relatively small, simple reaction mechanisms are easily handled with the basic or improved Euler methods. However, more complex mechanisms require more powerful tools. R has such tools, in the package `odesolve`, which stands for "ordinary differential equation solver". To access that package, you must install it in your R system either by using the package installer or by downloading from CRAN. Once it's installed, it will be there for subsequent use. To use it in a given R session, type `library(odesolve)`.

`odesolve` has two ODE solvers. The first, which we discuss in this section, is the 4th-order Runge-Kutta solver `rk4`, a standard method for solving initial value ODEs. We won't attempt to cover its derivation or details, but suffice it to say that its 4th-order nature enables it to efficiently solve a large range of problems that may involve rapidly varying functions. The other ODE solver is `lsoda`, which we will discuss in the next section. It is an "industrial strength" solver, well-suited to large sets of ODEs and those with widely varying time scales (so-called "stiff" equations).

Instructions for using `rk4` are given in the documentation for `odesolve`. (Type `?rk4` after `library(odesolve)`.) Since those instructions are rather terse, we adapt and elaborate the example given in the documentation to show how `rk4` is used. `rk4` must be given four arguments:

`yinit`, the vector of initial values for the dependent variables in the system of ODEs;

`times`, a vector of the times at which explicit estimates for `y` are desired, with the initial value of `times` as its first value;

`yprime`, the user-supplied function that computes the derivatives specified by the ODE. It must be called as `yprime = function(t, y, parms)`. `t` is the current time point in the integration, `y` is the current estimate of the variables in the ODE system, and `parms` is a vector of parameters. (The names of the function arguments can be different, so long as they are in the appropriate positions in the function call.) The return value of `yprime` should be a list, whose first element is a vector containing the derivatives of y with respect to time. If there is a second element, it should be a vector (possibly with a names attribute) of global values that are required at each point in `times`.

`parms`, a vector of parameters used in `func`. These parameters should be modifiable without rewriting the function, so that the solver can be used as part of a modeling package or for parameter estimation.

Then the routine is called as `rk4(yinit, times, yprime, parms)`. (There may be additional arguments, as described in the documentation, but we will not deal with them here.)

The example in the documentation is "a simple resource-limited Lotka-Volterra model" involving a substrate (s), a producer (p), and a consumer (k). The growth and consumption parameters are denoted by $a \ldots g$. To put the example into terms more familiar to the likely readers of this book, we consider the equivalent chemical reaction mechanism

$$S + P \xrightarrow{k_1} 2P$$
$$C \xrightarrow{k_2} S \qquad\qquad (6.19)$$
$$P + C \xrightarrow{k_3} 2C$$

which leads to the rate equations

$$\frac{d[S]}{dt} = -k_1[S][P] + k_2[C]$$
$$\frac{d[P]}{dt} = k_1[S][P] - k_3[C][P] \qquad\qquad (6.20)$$
$$\frac{d[C]}{dt} = k_3[C][P] - k_2[C]$$

where the ks are rate constants.

The code for solving the rate equations is then

```
> library(odesolve) # If odesolve is not already loaded
> # Reaction mechanism adapted from Lotka-Volterra-Model
> lvmodel = function(t, x, parms) {
    # x is the vector of concentrations
    # For convenience, give recognizable names to the
      components of x
    s = x[1] # Substrate
    p = x[2] # Producer
    c = x[3] # Consumer
    # Since parms has names, it must be called as a list
    with(as.list(parms), {
      ds = - k1*s*p + k2*c
      dp = k1*s*p  - k3*c*p
      dc = k3*p*c  - k2*c
      res<-c(ds, dp, dc)
      list(res)
    })
  }

> ## Vector of timesteps
```

```
> times   = seq(0, 100, length=101)

> ## Parameters for steady state conditions
> parms   = c(k1=0.1, k2=0.1, k3=0.1)

> ## Start values, with names for convenience
> xstart = c(s=2, p=1, c=1)

> ## Classical RK4 with fixed time step
> out1   = as.data.frame(rk4(xstart, times, lvmodel,
  parms))

> plot (out1$time, out1$s,  type="l", ylim=c(0,3))
> lines (out1$time, out1$p, type="l", lty=2)
> lines (out1$time, out1$c, type="l", lty=3)
```

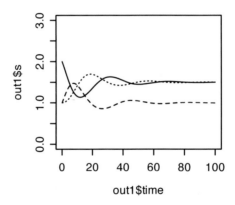

Fig. 6.10 Concentrations of S, C, and P in mechanism 6.19 calculated according to the 4th-order Runge-Kutta method

With the initial values of the concentrations of S, C, and P set at 1, and the values of the rate constants all equal, the system starts and remains in a steady state. (Verify this.) These restrictions can be changed in a variety of ways, as shown in the problems at the end of this chapter. One interesting change, used in the documentation, is to introduce a spike of S after a given time. The code uses the approx function, which returns a list of points that linearly interpolate given data points. (Note the similarity to the spline function.)

```
> lvmodel = function(t, x, parms) {
  s = x[1] # Substrate
```

```
   p = x[2] # Producer
   c = x[3] # Consumer
   with(as.list(parms),{
     import = approx(signal$times, signal$import, t)$y
     ds = - k1*s*p + k2*c + import
     dp = k1*s*p  - k3*c*p
     dc = k3*p*c  - k2*c
     res<-c(ds, dp, dc)
     list(res)
   })
}

> ## Vector of timesteps
> times   = seq(0, 100, length=101)

> ## External signal with rectangle impulse
> # Initially signal is set to zero everywhere
> signal = as.data.frame(list(times = times,
    import = rep(0,length(times))))
> # Then it is set to 0.2 between times 10 and 11
> signal$import[signal$times >= 10 &
    signal$times <=11] = 0.2

> ## Parameters for steady state conditions
> parms   = c(k1=0.1, k2=0.1, k3=0.1)

> ## Start values for steady state
> y<-xstart = c(s=1, p=1, c=1)

> ## Classical RK4 with fixed time step
> out1   = as.data.frame(rk4(xstart, times, lvmodel,
    parms))

> par(mfrow=c(1,2))
> plot (out1$time, out1$s,  type="l", ylim=c(0,3))
> lines (out1$time, out1$p, type="l", lty=2)
> lines (out1$time, out1$c, type="l", lty=3)
> # Phase plot of p vs c
> plot (out1$p, out1$c, type="l")
> par(mfrow=c(1,1))
```

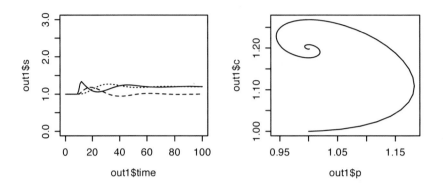

Fig. 6.11 (left) Concentrations of S, P, and C after a spike of S at $t = 10$. (right) Phase plot of C vs P

6.2.5 *lsoda, a sophisticated differential equation solver*

For very large systems of differential equations, or when many parameters need to be explored, `rk4` or Euler methods may be too slow. In that case, the method of choice is `lsoda`, which implements a compiled (and therefore very fast) Fortran routine behind the scenes. As R Help for `rk4` says, "The method is implemented primarily for didactic purposes. Please use lsoda for your real work!" `lsoda` should also be used for systems of stiff equations with widely varying time scales.

`lsoda` is called in the same way as `rk4`:
`lsoda(yinit, times, yprime, parms)`
If desired, relative and absolute error tolerances may be specified as additional arguments; see R Help.

Here is an example of a hypothetical kinetic mechanism that leads to oscillatory behavior [14, p. 203]

$$A \xrightarrow{k_1} X$$
$$X \xrightarrow{k_2} Y$$
$$X + 2Y \xrightarrow{k_3} 3Y \qquad\qquad (6.21)$$
$$Y \xrightarrow{k_4} R$$

This leads to the differential equations for the time dependence of X and Y, assuming A is constant:

$$\frac{d[X]}{dt} = k_1[A] - k_2[X] - k_3[X][Y]^2$$

$$\frac{d[Y]}{dt} = k_2[X] + k_3[X][Y]^2 - k_4[Y] \qquad (6.22)$$

R removes Y as a reactant from the system. We assume that [A] is a constant, a, and the initial concentrations of X and Y are zero. Code to solve these equations with lsoda and to plot X vs. t, Y vs. t, and X vs. Y (phase plot) is

```
> library(odesolve)
> # Specification of parameters
> k1 = .01; k2 = .01; k3 = 1e6; k4 = 1; a = .05
> # Parameter list to pass to function
> parms = c(k1, k2, k3, k4, a)
> # Initial concentrations of X and Y
> CX0 = 0; CY0 = 0
> # Time range and increment
> tmin = 0; tmax = 200; dt = 0.01
> # Vector of time values
> times = seq(tmin, tmax, bdt)

> # The function containing the derivatives of X and Y
> lsexamp = function(t, y, p)
    {
      # dy1 = dX/dt; dy2 = dY/dt
      dy1 = (p[1]*p[5]-p[2]*y[1]-p[3]*y[1]*y[2]^2)/p[4]
      dy2 = (p[2]*y[1]+p[3]*y[1]*y[2]^2-p[4]*y[2])/p[4]
      list(c(dy1, dy2)) # Output list
    }

# The lsoda routine now calls its various parts and
  sends the output to the array out
> out = lsoda(c(CX0,CY0),times,lsexamp, parms,
  rtol=1e-4, atol= 1e-6)

> # Dissecting the columns of out
> t = out[,1]; X = out[,2]; Y = out[,3];

> # Enabling three plots stacked vertically
> par(mfrow=c(3,1))

> plot(t,X,type="l") # X vs t
> plot(t,Y,type="l") # Y vs t
> plot(X,Y,type="l") # Phase plot

> par(mfrow=c(1,1)) # Reset the graphic parameters
```

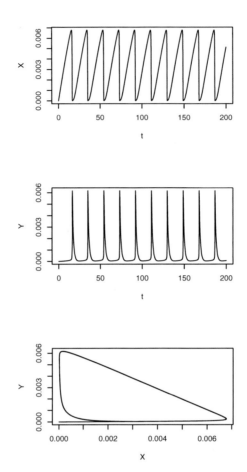

Fig. 6.12 Time evolution of X and Y, and phase plot of X vs Y, according to mechanism 6.21

6.2.6 Stability of systems of ODEs

Examination of the preceding mechanisms shows two different types of behavior: The Lotka-Volterra mechanism 6.19 converges to a steady state and the phase plot of S vs. P spirals to a point (an "attractor"); while the oscillating reactions mechanism 6.21 shows periodic repeats and the phase plot cycles. These two types of behavior can be predicted from analyses of the eigenvalues of the reaction matrices.

Let the reactants and products be components of the vector **x**, and the matrix of rate coefficients be **M**. Then we can symbolically represent the set of differential equations as

$$\frac{d\mathbf{x}}{dt} = \mathbf{M} \cdot \mathbf{x} \tag{6.23}$$

It can be shown that if the real parts of all of the eigenvalues of $\mathbf{M}$ are negative, the reaction converges to a stable equilibrium. If at least one of the real parts is positive, the reaction will not converge. If the eigenvalues have imaginary parts, oscillatory behavior is to be expected. These results hold for a linear reaction mechanism. If the mechanism involves higher-order powers of concentrations (as will be the case for bimolecular reactions), the results hold near equilibrium, where deviations from equilibrium are linear in the concentrations.

For example, consider the reaction

$$A + B \rightleftharpoons C \tag{6.24}$$

with forward and reverse rate constants k_+ and k_-, so the equilibrium constant is $K = k_+/k_-$. We denote concentrations by small letters and let $a = a_e + \Delta a$, $b = b_e + \Delta b$, and $c = c_e + \Delta c$, where subscript e denotes the equilibrium concentration. Then the rate of disappearance of A can be written

$$-\frac{d(a_e + \Delta a)}{dt} = k_+[(a_e + \Delta a)(b_e + \Delta b)] - k_-(c_e + \Delta c) \tag{6.25}$$

Since the equilibrium concentrations don't change with time ($da_e/dt = 0$) and the equilibrium concentrations and rate constants are related by $k_+ a_e b_e = k_- c_e$, eq. 6.25 is simplified to

$$-\frac{d\Delta a}{dt} = k_+(a_e \Delta b + b_e \Delta a + \Delta a \Delta b) - k_- \Delta c \tag{6.26}$$

But if we are close to equilibrium, the Δa and Δb terms are small, so their product is negligible. Therefore, we can finally write

$$-\frac{d\Delta a}{dt} = k_+(a_e \Delta b + b_e \Delta a) - k_- \Delta c \tag{6.27}$$

which is linear in the concentration deviations.

Now consider the Lotka-Volterra mechanism 6.19. Employing the linearization strategy, eqs. 6.20 become

$$\frac{d\Delta s}{dt} = -k_1(s_e \Delta p + p_e \Delta s) + k_2 \Delta c$$

$$\frac{d\Delta p}{dt} = k_1(s_e \Delta p + p_e \Delta s) - k_3(c_e \Delta p + p_e \Delta c) \tag{6.28}$$

$$\frac{d\Delta c}{dt} = k_3(c_e \Delta p + p_e \Delta c) - k_2 \Delta c$$

With the initial concentrations ($s = 2, p = 1, c = 1$) and rate constants given in the first rk4 example, we found that the equilibrium concentrations were $s_e = 1.5, p_e = 1.0, c_e = 1.5$. Thus the $\mathbf{M}$ matrix giving the coefficients of $\Delta s, \Delta p, \Delta c$ is

$$\mathbf{M} = \begin{pmatrix} -k_1 p_e & -k_1 s_e & k_2 \\ k_1 p_e & k_1 s_e - k_3 c_e & -k_3 p_e) \\ 0 & k_3 c_e & k_3 p_e - k_2 \end{pmatrix} = \begin{pmatrix} -1 & -1.5 & 1 \\ 1 & 0 & -1 \\ 0 & 1.5 & 0 \end{pmatrix} \qquad (6.29)$$

where we canceled out the rate constants since they are all equal. The eigenvalues of $\mathbf{M}$ are calculated from

```
> M=matrix(c(-1,-1.5,1,1,0,-1,0,1.5,0),nrow=3,byrow=T)
> eigen(M)$values
[1] -5.000e-01+1.658e+00i -5.000e-01-1.658e+00i
    -1.997e-16+0.000e+00i
```

There are three eigenvalues, one of which is zero. The other two have negative real parts, so the system will converge back to equilibrium if it is slightly perturbed. There are conjugate imaginary parts, which indicate that the reaction will show oscillations, as it does.

Applying the same approach to the oscillating reaction system, eq. 6.21, we find

$$\mathbf{M} = \begin{pmatrix} -k_2 - k_3 y_e^2 & -2k_3 x_e y_e \\ k_2 + k_3 y_e^2 & 2k_3 x_e y_e - k_4 \end{pmatrix} \qquad (6.30)$$

for the coefficients of Δx and Δy in the reaction matrix. We solve for the equilibrium concentrations using the equations, derivable from eqs. 6.22 and the linearization procedure

$$k_1 a - k_2 y_e - k_3 x_e y_e^2 = 0$$
$$k_2 x_e + k_3 x_e y_e^2 - k_4 y_e = 0 \qquad (6.31)$$

We solve these nonlinear equations numerically with the optim function and use the results to calculate the eigenvalues of $\mathbf{M}$.

```
> fr = function(x) {
  xe = x[1]; ye = x[2]
  k1 = .01; k2 = .01; k3 = 1e6; k4 = 1; A = .05
  (k1*A-k2*ye-k3*xe*ye^2)^2+(k2*xe+k3*xe*ye^2-k4*ye)^2
  }

> xe = optim(c(0,1e-4),fr)$par[1]
> ye = optim(c(0,1e-4),fr)$par[2]
> xe
[1] 0.001876
> ye
[1] 0.0005136
> k1 = .01; k2 = .01; k3 = 1e6; k4 = 1; A = .05
> M = matrix(c(-k2-k3*ye^2, -2*k3*xe*ye,k2+k3*ye^2,
  2*k3*xe*ye-k4),nrow=2,byrow=T)
> eigen(M)$values
[1] 0.3265+0.4089i 0.3265-0.4089i
```

The real parts of both eigenvalues are positive, so the system will not converge back to equilibrium if perturbed. Instead, as we have seen in the simulation, it oscillates in a stable pattern as indicated by the imaginary parts of the eigenvalues.

6.2.7 Numerical solution of second-order ODEs

The numerical methods (Euler, `rk4`, `lsoda`) that are used to solve systems of ODEs apply only to first-order systems; that is, systems in which equations for the first derivatives of the dependent variables are given. This is the type of ODE system encountered in chemical reactions, but there are other situations in biophysics where we encounter second-order (and sometimes higher order) equations.

Fortunately, it is easy to convert a second-order ODE to a system of two first-order equations. If the dependent variable is y, then a general second-order differential equation might be

$$\frac{d^2y}{dt^2} = f(y,t) \tag{6.32}$$

If we let $w = dy/dt$, then we have the two first-order equations

$$\frac{dy}{dt} = w \tag{6.33}$$

$$\frac{dw}{dt} = f(y,t) \tag{6.34}$$

6.3 Stochastic differential equations

Chemists are used to dealing with reactions involving an appreciable fraction of an Avogadro's number of molecules. With such a large number of molecules, our traditional approach of using deterministic rate equations is appropriate. However, within cells there may be only a handful of molecules of a given type, so the continuum approximation may no longer be satisfactory. Instead, a random or "stochastic" approach may be needed.

R has a package, `GillespieSSA`, for doing stochastic simulations. It implements various versions of the stochastic simulation algorithm (SSA) first developed by Gillespie [28]; see also the review by Gillespie [29]. The package, written by Mario Pineda-Krch, can be downloaded from the CRAN web site at `http://cran.r-project.org/`.

The original Gillespie algorithm simulated every successive molecular event, and is therefore usually too slow for practical simulation of realistic systems. "An approximate speedup to the SSA is provided by tau-leaping, in which time is advanced by a preselected amount τ and the numbers of firings of the individual reaction channels are approximated by Poisson random numbers" [29]. The `GillespieSSA` package implements both the direct method and three tau-leaping methods: Explicit tau-leaping (ETL), Binomial tau-leaping (BTL), and Optimized tau-leaping (OTL). All are called in a similar way, which at its simplest is

```
out = ssa(x0,a,nu,parms,tf,method)
```

where x_0 is a vector of the initial number of individuals of each type, **a** is a vector of "propensities" or probabilities a_j that a particular reaction j will occur in the next time interval, **nu** is a matrix whose elements v_{ij} are the changes in the number of individuals of type i caused by one reaction of type j, parms is a named vector of model parameters, tf is the final time, and method is one of "D", "ETL", "BTL", "OTL" with "D" for Direct as the default. GillespieSSA returns a list whose most important is data, a numerical matrix whose first column is the time vector and whose subsequent columns are the populations of the various types. See ?ssa in R Help for more details and options.

GillespieSSA comes with a number of demos including reaction, predator-prey, and infection models. In this chapter we'll use the Decaying-Dimerization Reaction Set as an example. This reaction scheme consists of three species and four reaction channels

$$S_1 \xrightarrow{k_1} 0$$

$$S_1 + S_1 \xrightarrow{k_2} S_2 \qquad\qquad (6.35)$$

$$S_2 \xrightarrow{k_3} S_1 + S_1$$

$$S_2 \xrightarrow{k_4} S_3$$

From this we see that the v matrix is

$$v = \begin{pmatrix} -1 & -2 & 2 & 0 \\ 0 & 1 & -1 & -1 \\ 0 & 0 & 0 & 1 \end{pmatrix} \qquad\qquad (6.36)$$

The propensity vector **a** is the product of the rate constants and populations for each of the reaction channels at a given time. We give an example using the direct method. By changing the method to one or more of the other three, you should be able to observe a substantial speed-up of the calculation. Note that the set.seed(1) command generates the same random numbers each time. This command should be commented out for work that intends to generate truly random results.

```
> library(GillespieSSA)
> # Define parameters
> parms = c(k1=1.0, k2=0.002, k3=0.5, k4=0.04)
> # Initial state vector
> x0 = c(s1=500, s2=0, s3=0)
> # State-change matrix
> nu = matrix(c(-1, -2, +2,  0,
                 0, +1, -1, -1,
```

```
                        0,    0,    0,  +1),
                        nrow=3,byrow=TRUE)
> # Propensity vector
> a   = c("k1*s1", "k2*s1*s1", "k3*s2", "k4*s2")
> tf = 30 # Final time

# Direct method
> # Generate the same random numbers each time
> set.seed(1)
> out = ssa(x0,a,nu,parms,tf,method="D")
> out.t = out$data[,1]
> out.s1 = out$data[,2]
> out.s2 = out$data[,3]
> out.s3 = out$data[,4]
> par(mfrow=c(1,3))
> plot(out.t,  out.s1, type="l")
> plot(out.t,  out.s2, type="l")
> plot(out.t,  out.s3, type="l")
> par(mfrow=c(1,1))
```

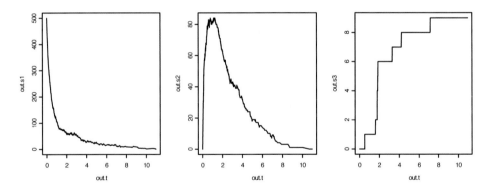

Fig. 6.13 Stochastic time evolution of the concentrations of S_1, S_2, and S_3 according to mechanism 6.35

6.4 Problems

1. Plot on the same graph the exact integration from 0 to 2 of $dy/dx = x^2$ (with $y(0) = 0$) and the numerical solution using Euler's method and $dx = 0.01$. Try the same calculation with $dx = 0.1$. How small does dx have to be for the exact and

numerical curves to be essentially indistinguishable? Does use of the improved Euler method enable a larger dx?

2. Repeat Problem 1 using `lsoda` from the `odesolve` package. What dx gives essentially exact numerical results?

3. Write the differential equations for the time-dependence of the concentrations of A, B, and C in the successive mechanism A $\rightarrow$ B $\rightarrow$ C with $k_1 = 10$ and $k_2 = 1$, and then with $k_1 = 1$ and $k_2 = 10$. Start with [A] = 10, [B] = [C] = 0. Plot the concentrations of A, B, and C on the same graph over the time 0–10, with suitable line colors and styles to distinguish the components and the two sets of rate constants. This problem illustrates the concept of "rate-limiting step".

4. Parallel reactions, symbolized by

can occur both chemically and in radioactive decay. For example, the unstable copper isotope $^{64}_{29}$Cu can decay into either $^{64}_{30}$Zn + β^- or $^{64}_{28}$Ni + β^+. The rate constants for decay into Zn and Ni are 0.0336 h^{-1} and 0.0206 h^{-1}, respectively. Write the rate equations and use the simple Euler method to solve for and plot the fractions of the three isotopes up to a time at which 90% of the Cu has decayed.

5. The Michaelis-Menten mechanism for enzyme kinetics is of the form

$$A + B \underset{k_{-1}}{\overset{k_1}{\rightleftharpoons}} X \underset{k_{-2}}{\overset{k_2}{\rightleftharpoons}} P + Q$$

where it is usually assumed that the first step is in rapid equilibrium and the second step is largely irreversible. What if this is not the case? Write the differential equations for the time-dependence of A, B, X, P, and Q, and solve them using the improved Euler method, with initial concentrations [A] = [B] = 1, [X] = [P] = [Q] = 0, and rate constants $k_1 = 3$, $k_{-1} = 1$, $k_2 = 2$, and $k_{-2} = 1$. Run the simulation until equilibrium is reached, and plot the concentrations of A, X, and P as functions of time. What are the equilibrium concentrations of the species?

6. Execute the code by which `lsoda` solves the oscillating chemical reaction model given in the example, and plot the results. Then repeat the simulation with $dt = 0.1$ and $dt = 1.0$. Comment on the behavior of the simulation as dt increases, and on how you would determine whether you've chosen a value adequate to get accurate simulation results. What problems might you encounter if you went to very small dt?

7. Another well-studied oscillating chemical reaction model is the *Brusselator*, so called because it was developed in Brussels by Ilya Prigogine and his collaborators. The mechanism [14, pp. 201–202] is

$$A \xrightarrow{k_1} X$$

$$B + X \xrightarrow{k_2} R + Y$$

$$Y + 2X \xrightarrow{k_3} 3X$$

$$X \xrightarrow{k_4} S$$

where A and B are reactants, R and S are products, and X and Y are interme-
diates. Write the differential equations for the rate of change of X and Y, and
solve them with `lsoda` as above over the time range from 0 to 250. Let the rate
constants be $k_1 = 0.025$, $k_2 = 1$, $k_3 = 1$, $k_4 = 0.01$, and vary the concentrations
of A and B in four "experiments":

(a) [A] = [B] = 0.01
(b) [A] = [B] = 0.02
(c) [A] = 0.01, [B] = 0.02
(d) [A] = 0.02, [B] = 0.01

Which starting concentrations lead to stable oscillations, and which to unstable?

8. Laidler (*J. Chem. Educ.* 49:343, 1972) showed an Arrhenius-type plot of ln frequency vs. $1/T$ of the frequency of chirping of tree crickets. The data are

```
recip.temp = c(3.334, 3.359, 3.381, 3.400, 3.422,
    3.444, 3.458, 3.480, 3.506)*1e-3
log.freq = c(2.298, 2.251, 2.201, 2.150, 2.106,
    2.061, 2.009, 1.963, 1.918)
```

Plot and analyze these data to get the apparent activation energy of the chirping
reaction.

9. The kinetics of very fast reactions cannot be studied by conventional mixing techniques, which are limited to milliseconds or slower. Instead, relaxation methods
are used, in which the equilibrium is perturbed, typically by a very short (μs to
ns) electric field or laser pulse to heat the solution, and then the rate of return
of equilibrium is followed. Since the method involves an abrupt increase in temperature, it is called the T-jump method. The perturbation from equilibrium is
usually small, so the process is usually first order and the return to equilibrium is
exponential.

An example is given in [62, pp. 369–370] of data from a 1972 study by Turner
et al. (*Nature* 239:215) who studied the kinetics of proflavin dimerization using
a laser T-jump method. For the dimerization reaction

$$2P \underset{k_{-1}}{\overset{k_1}{\rightleftharpoons}} P_2$$

the relaxation time τ (the time for an exponential to decay to e^{-1} of its initial
value) can be shown to relate to total P concentration according to the expression

$$\frac{1}{\tau^2} = k_{-1}^2 + 8k_1 k_{-1} [P]_{tot}$$

The following table gives data from this study, obtained by digitizing the graph in [62]. Data at pH 7.8 and pH 4.0 have been combined, since the reaction is not pH-dependent over this range.

Table 6.1 Data for Problem 9

$10^3 [P]_{tot}$, M	$10^{-12}/\tau^2$, s^{-2}
0.47	6.90
0.52	8.62
0.90	12.07
1.15	16.21
1.20	10.69
1.79	32.07
2.21	30.34
2.23	26.90
2.72	41.03
2.79	37.59
3.46	41.38
3.79	62.76
4.41	53.10
4.43	63.45
5.36	73.79

Save these data as a file on your hard drive, use `read.table()` to read the data into R, and plot and analyze them to get k_1, k_{-1}, and the equilibrium constant for dimerization $K = k_1/k_{-1}$. What are the standard errors of the estimated slope and intercept of the data, fit to a linear model? How do those carry over to standard errors in the estimated rate constants and equilibrium constant?

10. Exothermic reactions sometimes appear to catalyze themselves, because they liberate heat which makes the reaction go faster. The oscillating reaction discussed in the `odesolve` example and Problem 6 leads to interestingly complex behavior [14, pp. 204–206] if k_1 is made temperature-dependent according to the Arrhenius-type equation

$$k_1(T) = k_1(T_0) \exp(\theta)$$

where θ is a unitless activation energy. Then the differential equations to be solved become

$$\frac{d[X]}{dt} = k_1(T)[A] - k_2[X] - k_3[X][Y]^2$$

$$\frac{d[Y]}{dt} = k_2[X] + k_3[X][Y]^2 - k_4[Y]$$

$$\frac{d\theta}{dt} = \gamma[Y] - \beta\theta$$

along with the equation for $k_1(T)$. Solve these equations with lsoda over the time range from 0 to 300 using the parameters $k_1(T_0) = 1 \times 10^{-6}$, $k_2 = 5.5 \times 10^{-7}$, $k_3 = 1$, $k_4 = 1 \times 10^{-4}$, $\gamma = 10$, $\beta = 0.5$, $CX(0) = 0$, $CY(0) = 0$, $\theta_0 = 0$. Depending on the value of $[A] = a$ mol/l, you should get different types of behavior:

Table 6.2 Conditions for Problem 10

a	Oscillatory behavior
0.6000	period-1
0.6500	period-2
0.6870	period-4
0.6970	period-3
0.7080	chaotic

Run your program for each value of a, plot the results, and see what you get.

Chapter 7
Population Dynamics: Competition, Predation, and Infection

Populations—whether of organisms, cells, or viruses—are of central importance in biology. In this chapter we consider some of the basic models of population dynamics: competition of different species for resources, predation of one species upon another, and transitions of parts of a population between different states (e.g., susceptible, infected, resistant, dead) in epidemics.

Many models for the dynamics of populations have been developed. The simplest of these—which are the ones on which we will focus—are based on the models of reaction kinetics that we considered in Chapter 6. These simple models largely assume that organisms can be treated like molecules, in that the law of mass action applies, and that differences between organisms such as age and sex can be neglected. Despite these simplifications, the models are useful for many situations of interest to molecular biologists and biophysicists.

I have drawn heavily in this chapter on the compact survey of population and epidemic models by Kean and Spain [39], and on the extensive treatment of epidemics by Keeling and Rohani [38].

NOTE: Throughout this chapter, we use N as analogous to concentration in chemical systems: e.g., microbes per unit volume or organisms per unit area.

7.1 Models of homogeneous populations of organisms

7.1.1 Verhulst-Pearl (logistic) equation

Perhaps the simplest non-trivial population model is the logistic model developed by Verhulst in 1838 and rederived by Pearl and Reed in 1920. It recognizes that the exponential growth model

$$\frac{dN}{dt} = kN \tag{7.1}$$

V. Bloomfield, *Computer Simulation and Data Analysis in Molecular Biology and Biophysics,* Biological and Medical Physics, Biomedical Engineering, DOI: 10.1007/978-1-4419-0083-8_7, © Springer Science + Business Media, LLC 2009

is unrealistic except at very low population density, since as the population grows there must be a slowdown due to limited resources, competition among individuals, etc. In the simplest modification, k decreases linearly with N, and eventually reaches zero at saturation:

$$k = a - bN \Rightarrow \frac{dN}{dt} = N(a - bN) = aN - bN^2 \qquad (7.2)$$

This logistic equation produces an S-curve of growth: initially approximately exponential, then slowing as saturation begins, and finally stopping at saturation. It is frequently rewritten in the form

$$\frac{dN}{dt} = rN\left(1 - \frac{N}{K}\right) \qquad (7.3)$$

where r is the maximum rate and K is the maximum carrying capacity, or saturating population. Organisms are often classified as whether they are r-type (produce many more offspring than can possibly survive, e.g., many insect and fish species) or K-type (produce relatively few offspring, but nurture them to the maximum possible, e.g., humans and most mammals).

Equation 7.3 is the same as eq. 6.9 in Section 6.1.5 and has the same solution, though with somewhat different variables. Setting $K = L$ and $r = kL$ in eq. 6.10 , we see that the solution to eq. 7.3 can be written as

$$N(t) = \frac{K}{1 + (\frac{K}{N_0} - 1)e^{-rt}} \qquad (7.4)$$

At $t = 0$, $N = N_0$, and as $t \to \infty$, N approaches the carrying capacity K.

Although eq. 7.3 has an analytical solution, we can solve it numerically in R, plotting the solution along with the rate equation dN/dt and the per capita growth equation dN/Ndt as functions of N. We use an initial population $N_0 = 2$, a growth rate constant $r = 0.15$, and a carrying capacity $K = 100$. We also show the results if the population begins above the carrying capacity, at $N_0 = 120$.

```
> N0 = 2 # Start with 2
> r = 0.15; K = 100
> tmin = 0; tmax = 80; dt = 0.1
> t = seq(tmin,tmax,dt)
> n = (tmax-tmin)/dt + 1
> N = N0
> for (i in 2:n) {
dN = r*N[i-1]*(1-N[i-1]/K)*dt
N[i] = N[i-1]+dN
}

> N.120 = 120 # Start with 120, over carrying capacity
> for (i in 2:n) {
dN.120 = r*N.120[i-1]*(1-N.120[i-1]/K)*dt
N.120[i] = N.120[i-1]+dN.120
}
```

```
> dN.dt = r*N*(1-N/K)
> dN.Ndt = r*(1-N/K)
> dN.120.dt = r*N.120*(1-N.120/K)
> dN.120.Ndt = r*(1-N.120/K)

> par(mfrow=c(1,3))

> plot(t,N,type="l",ylim=c(0,120),main="Logistic Growth")
> lines(t,N.120,lty=2)
> legend(50,80,legend=c("N","N.120"),lty=1:2,bty="n")

> plot(N,dN.dt,type="l",ylab="dN/dt",main="Growth Rate",
xlim=c(0,max(N,N.120)),
ylim=c(min(dN.dt,dN.120.dt), max(dN.dt,dN.120.dt)))
> lines(N.120,dN.120.dt,lty=2)

> plot(N,dN.Ndt, type ="l",ylab = "dN/Ndt",
  main="Per Capita Growth Rate", xlim=c(0,max(N,N.120)),
ylim=c(min(dN.Ndt,dN.120.Ndt), max(dN.Ndt,dN.120.Ndt)))
> lines(N.120,dN.120.Ndt,lty=2)
> par(mfrow=c(1,1))
```

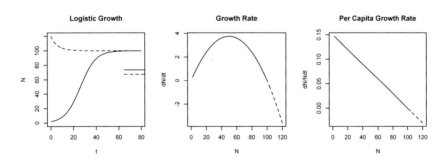

Fig. 7.1 Growth curve, growth rate, and per capita growth rate for logistic model

Check that the plot of growth rate vs. N has the expected maximum and intercept at $N = K$. In Problem 1 you will fit data on yeast growth to the logistic equation.

7.1.2 Variable carrying capacity and the logistic

In the treatment above it is assumed that the carrying capacity K of the environment is a constant, but this may not be the case. There may, for example, be seasonal fluctuations in temperature or daily fluctuations in light intensity. These may be approximated by making K fluctuate in a sinusoidal or sawtooth fashion. Here, for example, K varies sinusoidally with a period of 10.

```
> N0 = 2 # Start with 2
> r = 0.15
> tmin = 0; tmax = 80; dt = 0.1
> t = seq(tmin,tmax,dt)
> n = (tmax-tmin)/dt + 1
> # K has period 10, varies between 50 and 150
> K = 100 + 50*sin(2*pi*t/10)
> N = N0
> for (i in 2:n) {
dN = r*N[i-1]*(1-N[i-1]/K[i-1])*dt
N[i] = N[i-1]+dN
}
> plot(t,N,type="l",main="Oscillating carrying
  capacity",ylim=c(0,150))
> # Plot limits on carrying capacity
> abline(h=150,lty=3); abline(h=50,lty=3)
```

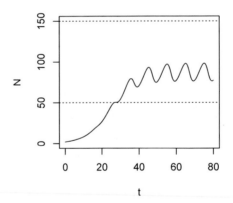

Fig. 7.2 Logistic growth under oscillating conditions

Note that the initial rise is unaffected by the oscillation in K, because it is below the lower limit. As the plateau is approached, the population stays well between the oscillating limits, because the response time (proportional to $1/r$) is not much less than the period of the oscillation.

7.1.3 Lotka-Volterra model of predation

A widely used model of predator-prey interaction is that due independently to Lotka (1926) and Volterra (1927). The model consists of two coupled equations for the dynamics of prey (N) and predator (P) populations:

$$\frac{dN}{dt} = k_1 N - k_2 NP \tag{7.5}$$

$$\frac{dP}{dt} = f k_2 NP - k_3 P \tag{7.6}$$

k_1 is the rate constant we have previously called r. k_2 is the rate constant for predation, while f is the efficiency of converting prey to predator biomass. Note that if there is no predation, the population of prey will grow exponentially. (This simplification will be removed in the next section.) The growth of the predator population depends entirely on its capture of prey. If this did not happen, P would decay exponentially to zero with rate constant k_3.

In calculating and graphing the dynamics of such coupled systems, it is useful not only to plot the individual populations vs. time, but also to plot one population vs. the other (P vs. N) in a phase plot. As the two populations oscillate, the phase plot will show a cycling behavior. The phase plot is divided into regions by isoclines, lines that represent equilibrium values for each population, and on either side of which the population either increases or decreases. The prey isocline is the horizontal line for which $dN/dt = 0$, i.e., $P = k_1/k_2$. The predator isocline is the vertical line for which $dP/dt = 0$, i.e., $N = k_3/f k_2$.

7.1.4 Modifications of the Lotka-Volterra model

Various modifications have been proposed to make the Lotka-Volterra model more realistic. In the first place, the unlimited increase in predation with increase in the prey population is unrealistic. A hyperbolic saturation model [34, 61]

$$\frac{dN}{dt} = k_1 N - \frac{k_2 NP}{K + N} \tag{7.7}$$

accounts for saturation with the factor K.

The Lotka-Volterra model can also be modified to include limits to growth of the prey and predator populations besides those imposed by each other. One way of doing this is shown in eqs. 7.8 and 7.9 from work of Leslie and Gower [41].

$$\frac{dN}{dt} = k_1 N \left(1 - \frac{N}{K} - \alpha P \right) \tag{7.8}$$

$$\frac{dP}{dt} = k_2 P \left(1 - \frac{P}{jN} \right) \tag{7.9}$$

Equation 7.8 reflects the limitation of prey growth by both carrying capacity K and predation $k_2 P$. Equation 7.9 states that when prey is abundant and predators scarce, the predator population grows at the maximum rate $k_2 P$; while as prey becomes scarce or predators abundant, the growth rate is reduced or made negative by the factor P/jN.

An equation for the growth rate of prey that takes into account both the Hollings-Tanner and Leslie-Gower modifications is

$$\frac{dN}{dt} = k_1 N \left(1 - \frac{N}{K_1} - \frac{\alpha P}{K_2 + N} \right) \tag{7.10}$$

7.1.5 Volterra's model for two-species competition

Another important type of population dynamics is when two species inhabit the same territory and consume the same resources, albeit with different efficiencies. Logistic-type equations for this model of competition are

$$\frac{dN_1}{dt} = k_1 N_1 \left(1 - \frac{N_1 + \alpha N_2}{K_1} \right) \tag{7.11}$$

$$\frac{dN_2}{dt} = k_2 N_2 \left(1 - \frac{N_2 + \beta N_1}{K_2} \right) \tag{7.12}$$

where α and β are the competition coefficients. The isoclines for species 1 and 2 are diagonals on an N_2 vs. N_1 phase plot:

$$N_1 = K_1 - \alpha N_2 \tag{7.13}$$
$$N_2 = K_2 - \beta N_1 \tag{7.14}$$

Depending on whether and how the two isoclines cross (that is, depending on the parameters in parentheses), four outcomes are possible: either species may "win" (exterminate the other) or there may be a stable or unstable equilibrium between them.

7.2 Models of microbial growth

Some models other than those we've considered up to now have commonly been used for growth and competition of microbes. This is due in part to real differences in behavior of these simple microorganisms, and in part to different growth conditions (batch or continuous) that affect the supply of nutrients.

7.2.1 Monod model of microbial growth

Jacques Monod, one of the founders of molecular biology, observed that the assimilation of nutrients into the biomass of bacteria followed the same hyperbolic mathematical form as that of enzyme kinetics: linear at low substrate, and leveling off at saturating substrate.

$$\frac{dB}{dt} = \mu B = \mu_m \frac{[S]}{K_s + [S]} B \tag{7.15}$$

B is cell density, $[S]$ is the nutrient concentration, μ is the growth rate constant, μ_m is the maximum growth rate (analogous to V_{max}), and K_s is the half-saturation nutrient concentration (analogous to K_m). There may be losses in microbial density because of respiration, mortality, and autolysis, characterized by a rate constant R, so a modified growth equation is

$$\frac{dB}{dt} = \mu B = \left(\mu_m \frac{[S]}{K_s + [S]} - R \right) B \tag{7.16}$$

7.2.2 Microbial growth in batch culture and chemostat

In most laboratories the standard way to grow microbes is in batch culture. A small number of microbes are added to a container of nutrient substrate medium, and the population grows until the substrate is exhausted. As the microbes die, they may not release the nutrients needed to fuel the reproduction of new organisms, so the population will tend to decline from the level set by the initial carrying capacity. The total concentration of substrate is given by

$$[S]_{tot} = [S] + aB \tag{7.17}$$

where a is the proportion of cell biomass made from substrate.

A better-controlled way to study the population dynamics of microorganisms under various experimental conditions is to use a chemostat [57]. The microbes are put into a vessel of volume V containing well-stirred liquid nutrient medium, into which new medium flows at rate F (volume/time), and from which medium is removed at the same rate. The removed medium contains living and dead cells, unused medium, and metabolic products. We denote the cell biomass density by B, the concentration of nutrient substrate flowing into the chemostat by $[S_i]$, the concentration of substrate in the chemostat and outflow by $[S]$, and the dilution rate F/V (1/time) by D. Then the rates of change of cell biomass and substrate concentration in the chemostat are

$$\frac{dB}{dt} = \mu B - DB \tag{7.18}$$

$$\frac{d[S]}{dt} = D]S_i] - D[S] - \mu aB \tag{7.19}$$

7.2.3 Multiple limiting nutrients

Most organisms require more than one nutrient to grow, and which one is limiting may depend on the particular situation. For example, diatoms require silicon (S), nitrogen (N), and phosphorus (P). An adaptation of the Monod equation becomes

$$\mu = \mu_m \left(\frac{S}{K_S + S} \right) \left(\frac{N}{K_N + N} \right) \left(\frac{P}{K_P + P} \right) \tag{7.20}$$

where we have omitted the square brackets that denote concentration for notational simplicity. There is also an equation of the form $[S]_{tot} = [S] + aB$ for each nutrient.

7.2.4 Competition for limiting nutrients

A pertinent example is the competition of green algae and blue-green algae for nitrogen and phosphorus in lakes. Some blue-green algae can fix molecular nitrogen, using atmospheric nitrogen dissolved in water; green algae cannot use this source of nitrogen. The original paper [51] considered a variety of factors affecting nutrient dynamics, including "runoff, mixing, sinking of algae, [light and temperature variation,] and predation by herbivorous zooplankton" [39]. However, we shall consider only competition for N and P.

The basic equations for this model, with G standing for green algae, B for blue-green algae, n for dissolved molecular nitrogen gas, and other variables whose meanings should be familiar by now, are

$$\frac{dG}{dt} = \mu_G G - R_G G, \quad \mu_G = \mu_{mG} \left(\frac{[P]}{K_{pG} + [P]} \right) \left(\frac{[N]}{K_{nG} + [N]} \right) \tag{7.21}$$

$$\frac{dB}{dt} = \mu_B B - R_B B, \quad \mu_B = \mu_{mB} \left(\frac{[P]}{K_{pB} + [P]} \right) \left(\frac{[N] + [n]}{K_{nB} + [N] + [n]} \right) \tag{7.22}$$

Algal death and respiration will release P and N to the lake water. Therefore, the available phosphorus concentration will be

$$[P] = [P]_{tot} - a_{pG} G - a_{pB} B \tag{7.23}$$

where $[P]_{tot}$ is the concentration of total P, and a_{pG} and a_{pB} are the fractions of P in the biomass of the two algal species.

The total nitrogen concentration, $[N]_{tot}$, will increase due to nitrogen fixation as B increases, and will decrease due to denitrification—the bacterial breakdown of nitrogenous compounds to molecular nitrogen. The equations for these processes and then for the resulting N concentration are

$$\frac{d[N]_{tot}}{dt} = -k_D[N] + a_{nB}B\mu_{mB}\left(\frac{[N]+[n]}{K_{nB}+[N]+[n]} - \frac{[N]}{K_{nB}+[N]}\right) \tag{7.24}$$

$$[N] = [N]_{tot} - a_{nG}G - a_{nB}B \tag{7.25}$$

where k_D is the dentrification rate constant. Most lakes are saturated with molecular nitrogen, so $[n]$ may be considered a constant.

7.2.5 Toxic inhibition of microbial growth

The effects of a toxic compound T on microbial growth can be modeled by equations similar to those for enzyme inhibition: either noncompetitive inhibition, in which even high concentrations of nutrient S cannot restore the growth rate

$$\mu = \mu_m\left(\frac{1}{1+[T]/K_T}\right)\left(\frac{[S]}{K_s+[S]}\right) \tag{7.26}$$

or competitive inhibition, in which T increases the concentration of S required to reach a given growth rate:

$$\mu = \mu_m\left[\frac{[S]}{[S](K_s(1+[T]/K_T)}\right] \tag{7.27}$$

7.3 Models of Epidemics

Modeling of infectious diseases is one of the most practical and important applications of computer simulations in biology. Modeling can have a variety of purposes, e.g., predicting the outcome of an epidemic, understanding the underlying biological and epidemiological processes, or providing guides to public health and safety issues. The level of detail in an epidemiological model should reflect its purpose. General issues of epidemiological modeling, as well as specific models approached from a computational point of view, are cogently treated by Keeling and Rohani [38], and at a more mathematical level by Diekmann and Heesterbeek [19]. Chapter 24 in Keen and Spain [39] also has valuable material. The book *Virus Dynamics* by Nowak and May [48] uses epidemiological concepts to consider the dynamics of HIV infection, where the population under attack is the cells of the immune system.

Diseases can be classified as infectious (e.g., influenza) and noninfectious (e.g., cancer), and the agents of infectious diseases can be either microparasites (e.g., viruses) or macroparasites (e.g., worms). A further distinction is whether the infection is passed by direct contact between individuals, or indirectly, in which case the parasite spends some part of its life cycle outside the host organism. We shall consider epidemic models of directly transmitted infections caused by microparasites.

A common process of infection of a host organism can be considered to occur in four stages: susceptible (S), in which no pathogen is present; exposed (E), in which there are no symptoms and a level of pathogen too low to be further transmitted; infectious (I), in which the pathogen level is high enough to transmit to additional hosts, whether or not illness is overt; and recovered (R), in which the host's immune system has cleared the microparasite and the host is no longer infectious. (R may also stand for "removed", if the host is quarantined or is killed by the infection.) This four-stage disease model is often abbreviated SEIR. Frequently, the exposed stage is omitted from the model equations, so an SIR model is often used.

Standard responses to epidemics affect the populations of these stages in different ways. Vaccination reduces the number of susceptibles. Quarantine reduces the number of infected individuals mixing with the general population. Culling of animal herds reduces the numbers of both susceptible and infectious hosts; modeling may indicate whether this will lead to a lower total number of deaths than letting the infection take its course.

Models other than simple SIR may also be pertinent. For example, infection by a sexually transmitted disease such as gonorrhea typically neither kills nor immunizes the host against further infection, so obeys an SIS model. More detailed models may deal with age or sex heterogeneity in the population, multiple pathogens or multiple hosts (e.g., transmission of influenza from birds to humans), seasonal behavior, spatial effects in geographically separated regions, and stochastic effects in small populations.

7.3.1 The simple SIR model

We consider first the SIR model in a "closed system" without the complications of birth, death, or migration. We let S, I, and R be the proportions of the population in the three states, so that

$$S + I + R = 1 \qquad (7.28)$$

Assuming a mass action mechanism for the infection of susceptibles by infectious members of the population, we have three differential equations governing the rates of change of the three groups:

$$\frac{dS}{dt} = -\beta SI \tag{7.29}$$

$$\frac{dI}{dt} = \beta SI - \gamma I \tag{7.30}$$

$$\frac{dR}{dt} = \gamma I \tag{7.31}$$

β, which functions as a rate constant, is the product of contact rate and transmission probability; and γ is the recovery or removal rate constant. One of these equations (usually eq. 7.31) can be omitted because of the population conservation equation 7.28.

In Problem 12 you are asked to numerically solve these equations for a particular set of parameters. However, some important insight can be gained analytically. Equation 7.30 can be rewritten

$$\frac{dI}{dt} = I(\beta S - \gamma) \tag{7.32}$$

If the initial fraction of susceptibles, $S(0)$, is less than γ/β, then $dI/dt < 0$ and the infection will die out. This important "threshold" result was derived by Kermack and McKendrick [40], the developers of SIR theory. Another way to look at this is that the relative removal rate γ/β must be small enough to allow propagation of the disease.

The inverse of the relative removal rate, β/γ, is called the *basic reproductive rate* symbolized by R_0. It represents (on average in any real situation) the number of secondary infections arising from a single primary infection in an entirely susceptible population ($S_0 = 1$). R_0, a key parameter in the theory of epidemics, must be > 1 if a pathogen is to successfully invade an initially completely susceptible population. Values of R_0 for some familiar diseases run from 2.44 for rabies in dogs in Kenya to 16–18 for measles or whooping cough in humans in the UK [38, Table 2.1].

Further insight can be obtained by dividing eq. 7.29 by eq. 7.31 to get

$$\frac{dS}{dR} = \frac{-\beta S}{\gamma} = -R_0 S \tag{7.33}$$

which can be integrated (assuming $R(0) = 0$) to give

$$S(t) = S(0)e^{-R_0 R(t)} \tag{7.34}$$

(Don't confuse the basic reproduction ratio R_0 with the fraction of recovered at time t, $R(t)$.) Since $R \leq 1$, S must always be greater than e^{-R_0} and so can never decline to zero. In other words, there will always be some susceptibles that avoid infection. Thus "[t]he chain of transmission eventually breaks down due to the decline in infectives, *not* due to a complete lack of susceptibles" [38].

7.3.2 The SIR model with births and deaths

We have seen that the infection will eventually die out in the SIR model, but this is not what is generally observed in real life. A more realistic result is obtained if we consider that the pool of susceptibles is continually renewed by births, while the population is kept constant by deaths at the same rate. (The deaths are not due to the epidemic; S, I, and R all die at the same rate.) If the average lifetime is 1μ, the death rate is μ and we assume that this is also the birth rate. Then the SIR equations become

$$\frac{dS}{dt} = \mu - \beta SI - \mu S \qquad (7.35)$$

$$\frac{dI}{dt} = \beta SI - \gamma I - \mu I \qquad (7.36)$$

$$\frac{dR}{dt} = \gamma I - \mu R \qquad (7.37)$$

From eq. 7.36 we see that the average time spent in the infectious class is $1/(\gamma + \mu)$, while the rate of infection is β. When the entire population is susceptible, so that $S = 1$, the basic reproduction ratio is the product of these two factors

$$R_0 = \frac{\beta}{\gamma + \mu} \qquad (7.38)$$

This is less than R_0 for a closed population because deaths lower the average time that an individual is infectious.

Equations 7.35–7.37 can each be set equal to zero to solve for the equilibrium values of the three classes in the population. After some algebra, we find [38, p. 29]

$$S_{eq} = \frac{1}{R_0} \qquad (7.39)$$

$$I_{eq} = \frac{\mu}{\beta}(R_0 - 1) \qquad (7.40)$$

$$R_{eq} = 1 - \frac{1}{R_0} - \frac{\mu}{\beta}(R_0 - 1) \qquad (7.41)$$

These values hold, and the equilibrium is stable, only if $R_0 > 1$, as otherwise the equilibrium values of I and R will be zero.

As numerical solution of eqs. 7.35–7.37 will demonstrate (see Problem 14), the approach to equilibrium occurs via damped oscillations.

7.3.3 SI model of fatal infections

The foregoing treatment has assumed that infection does not cause death; the I and R populations have normal life spans. There are some diseases that are invariably fatal, however. For those diseases, there is no R class, only S and I. If we let X be the number in the S class, and Y be the number in the I class, so the total population is $X + Y$, and assume that the transmission rate depends on the population size (rather than being independent of total population as in previous examples), then we have the two simultaneous differential equations

$$\frac{dX}{dt} = v - \beta XY - \mu X \tag{7.42}$$

$$\frac{dY}{dt} = \beta XY - (\gamma + \mu)Y \tag{7.43}$$

where v is the birth rate (assumed constant), β is the transmission rate, $1/\gamma$ is the average time that individuals remain infectious before they die, and μ is the normal death rate.

It can be shown [38, p. 39] that there is a stable equilibrium with

$$X_{eq} = \frac{\gamma + \mu}{\beta} \tag{7.44}$$

$$Y_{eq} = \frac{v}{\gamma + \mu} - \frac{\mu}{\beta} \tag{7.45}$$

so long as $R_0 = \frac{\beta \, nu}{(\mu + \gamma)\mu} > 1$.

7.3.4 SIS model of infection without immunity

We have previously considered epidemics in which hosts, once infected, either develop immunity (become R) or die. Another possibility, and a common one in real life, is that infection confers no lasting immunity, so that infected individuals return to the susceptible pool. One familiar example is the propagation of the common cold in a school classroom; another is the flow of a sexually transmitted disease such as gonorrhea, which we will consider in the next section.

For the SIS model, there is no R class and $S + I = 1$. The differential equations are

$$\frac{dS}{dt} = \gamma I - \beta SI \tag{7.46}$$

$$\frac{dI}{dt} = \beta SI - \gamma I \tag{7.47}$$

where we have ignored births and deaths. Since $S = 1 - I$, we can substitute for S in eq. 7.47 and obtain, after some rearrangement,

$$\frac{dI}{dt} = \beta I[(1 - 1/R_0) - I] \tag{7.48}$$

where $R_0 = \beta/\gamma$. Equation 7.48 has the same form as the logistic equation and can be integrated to give the closed-form solution

$$I(t) = \frac{I_0(1 - 1/R_0)}{I_0 + (1 - 1/R_0 - I_0)e^{-\beta(1 - 1/R_0)t}} \tag{7.49}$$

7.3.5 A model for an epidemic of gonorrhea

The SI model for epidemics can be used to simulate a gonorrhea epidemic in human populations.

Gonorrhea is caused by the bacterium *Neisseria gonorrhoeae*, which is spread almost exclusively by sexual contact. Many carriers show no external symptoms. Gonorrhea can be treated with antibiotics, but previous infections do not immunize against further infections.

Promiscuous males (number N_M) and females (number N_F) must be modeled as two separate but interacting populations. Dividing both populations into susceptible and infectious, we have

$$N_M = S_M + I_M \tag{7.50}$$
$$N_F = S_F + I_F \tag{7.51}$$

The rate of infection of susceptible males, assuming mass action behavior, will be $k_1 I_F S_M$, and of susceptible females $k_3 I_M S_F$. k_1 and k_3 are the infection coefficients, with units of (male or female)$^{-1}$ time^{-1}. Homosexual transmission is assumed to be insignificant.

The infectious carriers may seek treatment for the disease, and successful treatment will change an infectious to a susceptible. Treatment rates are $k_2 I_M$ and $k_4 I_F$; the rate constants have units of time^{-1}. Usually k_2 will be larger than k_4 because infected males develop painful symptoms which encourage them to seek treatment, while symptoms in females may be less severe.

The change in numbers of infectious carriers is obtained by combining the expressions for gain by infection and loss by treatment.

$$\frac{dI_M}{dt} = k_1 S_M I_F - k_3 I_M \tag{7.52}$$

$$\frac{dI_F}{dt} = k_3 S_F I_M - k_4 I_F \tag{7.53}$$

Problem 17 asks you to simulate the behavior of a gonorrhea infection with empirical values for the parameters. See [10, p. 465] for a detailed treatment of this model.

7.4 Problems

1. As an example exercise given by Keen and Spain [39, p. 107], Pearl [50] collected these data (mg biomass/100 ml culture) from a culture of yeast cells:

```
hours = c(0,1,2,3,4,5,6,7,8,9,10,12,14,18)
yeast = c(4,7,12,19,28,48,70,103,140,176,205,238,256,265)
```

Use the data fitting techniques previously discussed to estimate r and K, then use these parameters to plot the growth curve as a continuous curve with Pearl's data points as circles. According to Keen and Spain, it will be necessary to use a small increment in t (e.g., $\Delta t = 0.0001$) to achieve adequate convergence of the calculation, or a larger value, $\Delta t = 0.1$), with the improved Euler method (Section 6.2.3).

2. Write and run a computer simulation of the Lotka-Volterra model, showing two graphs: N and P vs. t, and a phase plot of P vs. N with isoclines. Use the two-stage improved Euler method with $\Delta t = 0.1$ or less. Use the parameters $k_1 = 0.1$, $k_2 = 0.002$, $k_3 = 0.2$, $f = 0.1$, $N_0 = 1500$, $P_0 = 50$. Allow the simulation to proceed for at least four complete predator-prey cycles. You may need to multiply predator or prey density by some constant to clearly show fluctuations on the first graph.

3. Assume that the prey is an insect that attacks an agriculturally valuable crop, and that its predator is therefore an agriculturally beneficial insect. Modify the previous simulation to add a pesticide that reduces the prey density to 55% of its pre-pesticide value near the *maximum* in the prey population in the second cycle. Repeat this simulation, but add the pesticide near the *minimum* in the prey population. This will show that the timing of pesticide application can have significant impact on predator-prey dynamics.

4. Numerically solve the set of equations 7.10 and 7.9. Use the parameters $k_1 = 0.4$, $K_1 = 900$, $\alpha = 2.5$, $K_2 = 200$, $k_2 = 0.1$, $j = 0.5$, $N_0 = 25$, $P_0 = 5$, $\Delta t = 0.1$ with the improved Euler method. Plot N and P vs. t on the same graph, and run the simulation until the populations have reached their steady states.

5. Solve these equations for the predator and prey isoclines, and graph the isoclines on a phase plot of prey density (x-axis) vs. predator density (y-axis). First use the parameters above, then rerun the simulation using $k_1 = 0.3$, $K_1 = 1000$, $\alpha = 1$, $K_2 = 200$, $k_2 = 0.03$, $j = 0.45$, $N_0 = 60$, $P_0 = 10$.

6. Write and run an R simulation for the Volterra competition model 7.1.5, beginning with $N_1 = 10$, $N_2 = 10$ and using the parameters $k_1 = 0.8$, $k_2 = 1.0$, $K_1 = 300$, $K_2 = 300$, $\alpha = 0.5$, $\beta = 0.6$. Use two-stage simple Euler integration, allow the simulation to run to equilibrium (about 25-30 time units), and graph the

results on a phase plot. Determine which outcome this set of parameters leads to. Then modify the parameters to obtain the other three outcomes.

7. Use eq. 7.16 to simulate bacterial growth in a batch culture with limiting nutrient, where the total nutrient is apportioned between free nutrient and bacterial biomass according to $[S]_{tot} = [S] + aB$ where a is the fraction of biomass concentration B made up of absorbed nutrient. Use the improved Euler method to run the simulation with the following values for the parameters—$R = 0.03$/hr, $a = 0.03$, $\mu_m = 0.3$/hr, $[S]_{tot} = 100$ mg/liter, $K_s = 25$ mg/liter,$B_0 = 1$ mg/liter— until a steady state is reached. Adjust the initial value of $[S]$ to include the nutrient in the initial inoculum. Plot B and $[S]$ vs. t on the same scale.

8. Simulate growth of a microorganism in a chemostat using eq. 7.15, eq. 7.19, and the parameters $[S_i] = 100$ mg/liter, $B_0 = 10$, $[S]_0 = [S_i]$, $K_s = 75$, $\mu_m = 1.5$, $a = 0.013$. Calculate the effects on B and $[S]$ of four different dilution rates—$D = 0.05, 0.25, 0.50$, and 0.70—and plot separate graphs for B and $[S]$ over 50 units of time. Use simple two-stage Euler integration with $\Delta t = 0.1$.

9. Write and run an R program to simulate multiple nutrient limitation of diatom growth in a batch culture. Use the following parameters:

Table 7.1 Data for PopMod-13

	S	N	P
Total nutrient concs, mg/liter	10.0	1.0	0.1
Fractional comp. of diatom biomass (a)	0.15	0.10	0.005
Half-saturation consts, mg/liter	0.10	0.15	0.04

Initial diatom biomass: $B = 0.10$ mg/liter
Rate of loss for mortality, etc.: $R = 0.005$ hr^{-1}
Maximum growth rate at saturation: $\mu_m = 0.120$ $^{-1}$
Use two-stage simple Euler integration, with $\Delta t = 1$, and let the simulation proceed for 120 hours. Plot B, S, N, and P vs. t on the same graph. Which nutrient is limiting under these conditions?
Then repeat this simulation with 100 times the concentration of each of the baseline nutrient concentrations in turn. You should discover that one of the nutrients limits the final level of B, but another limits the rate of growth.

10. Write and run an R program to simulate the competition between green and blue-green algae for limiting P and N. Use the following parameters:
$K_{pG} = 0.05$ mg/liter, $K_{pB} = 0.03$ mg/liter
$K_{nG} = 0.30$ mg/liter, $K_{nB} = 0.20$ mg/liter
$a_{pG} = 0.01$ mg P/mg biomass, $a_{pB} = 0.01$ mg P/mg biomass
$a_{nG} = 0.08$ mg N/mg biomass, $a_{nB} = 0.08$ mg N/mg biomass
$\mu_{mG} = 2.0$/day, $\mu_{mB} = 1.0$/day
$R_G = 0.06$/day, $R_B = 0.04$/day
$k_D = 0.05$/day
Constant concentrations: $[n] = 0.02$ mg/liter, $[P]_{tot} = 0.05$ mg/liter
Initial variable concentrations (mg/liter): $G = 0.01$, $B = 0.01$, $[N]_{tot} = 0.1$

Use two-stage simple Euler integration with $\Delta t = 1.0$ day, and run the simulation for 240 days. Plot $[N]_{tot}$ and the concentrations of B, G, and available N over this time period.

11. Select either the competitive or noncompetitive equation to simulate the effect of adding a toxicant to a batch microbial culture. Use the following parameters: $\mu_m = 1$, $K_s = 25$, $R = 0.03$, $a = 0.03$, $K_T = 0.01$, initial B concentration $= 1$ mg/liter.
    Make your simulation in two parts:

    a. Plot dB/dt vs. [T] from 0 mg/liter to 0.04 mg/liter, in increments of 0.001 mg/liter. Do this for at least 4 different values of [S]. These values should cover a fairly wide range, including much less than K_s, about equal to K_s, greater than K_s, and much greater than K_s to simulate saturation.
    b. Plot the growth of the culture through time with at least six different concentrations of T: 0, 0.001, 0.01, 0.1, 1, and 10.

12. Numerically solve the SIR model equations for an initially entirely susceptible population with one infectious individual and a total population of 1000. Use $\beta = 1.5$ per day and $1/\gamma = 7$ days. Plot S, I, and R out to 10 weeks.

13. (a) Plot the "epidemic curve", the number of new cases per unit time $= -dS/dt$, for the conditions in the previous problem. (b) Calculate and plot the total fraction of the population ultimately infected in the SIR model as the basic reproduction ratio R_0 varies from 1 to 5.

14. (From [38, p. 32]) Numerically solve eqs. 7.35–7.37 with the parameters $1/\mu = 70$ years, $\beta = 520$/year, and $1/\gamma = 7$ days. Start with the nonequilibrium values $S(0) = 0.1$ and $I(0) = 2.5 \times 10^{-4}$, and plot I vs. time for 70 years.

15. Calculate and plot the time dependence of X and Y for the SI model equations 7.42 and 7.43 with $X(0) = 10^4$, $Y(0) = 1$, and values of ν, μ, β, and γ chosen so that $R_0 > 1$.

16. Equation 7.49 shows that the fractions of infectious and susceptible in the population depends on the initial infectious fraction for the SIS model. Plot the time-dependence of the susceptible fraction for a population that starts with one infectious individual and 20, 50, and 100 total individuals. How does the decrease in S depend on total population size?

17. Write an R program based on eqs. 7.50–7.53 to simulate the course of an epidemic of gonorrhea with promiscuous groups of 900 males and 600 females. According to data given in [39], plausible daily rate constants are $k_1 = 0.000032$, $k_2 = 0.2$, $k_3 = 0.00033$, $k_4 = 0.025$. Start the simulation with one infectious male and no infected females. Calculate and plot the numbers of infected males and females over a 12-year (4380 day) period. Does the simulation eventually converge to a steady state?

Chapter 8
Diffusion and Transport

The movement of biological molecules in cells or in lab experiments gives useful insight into their sizes, associations, and mechanisms of transport to functional locations. The movement may be random diffusion (Brownian motion), it may be in response to some driving force, or both. Familiar driving forces in the lab are electrophoretic and centrifugal fields. In the cell, active transport and transport by cytoskeletal fibers are important mechanisms. Related situations arise in drug delivery, where the flow of drug from one compartment of the body to another can be treated by diffusion and transport models. Diffusion may be coupled with reaction, as discussed in Chapter 10, Section 10.3, on regulation of morphogenesis. In this chapter we will develop simple simulations for some of these processes.

8.1 Transport by simple diffusion

Diffusion can be viewed at the macroscopic level as the net flow of molecules due to a concentration gradient from high to low concentration, or at the microscopic level as the random thermal motion of molecules. If there are more molecules in one region than another, random motion will drive more molecules from the high-concentration region to the low than in reverse, so the two descriptions are equivalent.

8.1.1 Fick's laws of diffusion

We begin with the macroscopic picture, which is described by Fick's first law of diffusion:

$$J = -D\frac{\partial c}{\partial x} = -D\frac{\Delta c}{\Delta x} \tag{8.1}$$

V. Bloomfield, *Computer Simulation and Data Analysis in Molecular Biology and Biophysics,*   159
Biological and Medical Physics, Biomedical Engineering, DOI: 10.1007/978-1-4419-0083-8_8,
© Springer Science + Business Media, LLC 2009

where J is the flow of material per second across a unit area perpendicular to the direction of flow (visualize a (semi)permeable membrane of unit area separating the two regions of concentration); c is the concentration, x is the direction of flow, and D is the diffusion coefficient (cm^2/sec in cgs units or m^2/sec in SI units). J is measured in the same concentration units as c: if c is in g/cm^3, J is in $g/(cm^2$-sec). The partial derivative in the middle equality indicates that c may vary with time as well as with x-position, and also perhaps in the y- and z-directions. In our simple treatment here, however, we will begin with only one-dimensional diffusion. The second equality replaces the differentials with finite differences, which is the form needed to set up the simulation.

As flow proceeds, the concentration and flow at each point in the fluid will change. Consider a volume element in the fluid, with cross-sectional area A, thickness dx in the direction of flow, and volume $dV = Adx$. The flow during time dt into the volume element at x is $AJ(x)dt$ and the flow out at $x+dx$ is $AJ(x+dx)dt$. The difference between inward and outward flows is the amount of material dQ that accumulates in the volume element:

$$dQ = dc(x,t)dV = [J(x,t) - J(x+dx,t)]Adxdt = -\left(\frac{\partial J}{\partial x}\right)_t dVdt \qquad (8.2)$$

where the last equality comes from a Taylor's series expansion of J around x. It expresses the conservation of matter during the diffusion process. Rearranging, we find

$$\left(\frac{\partial c}{\partial t}\right)_x = -\left(\frac{\partial J}{\partial x}\right)_t \qquad (8.3)$$

Substituting Fick's first law, eq. 8.1, we obtain Fick's second law of diffusion

$$\left(\frac{\partial c}{\partial t}\right)_x = \frac{\partial}{\partial x}\left(D\frac{\partial c}{\partial x}\right)_t \qquad (8.4)$$

If D is independent of position, as is often assumed to be the case, eq. 8.4 becomes

$$\left(\frac{\partial c}{\partial t}\right)_x = D\left(\frac{\partial^2 c}{\partial x^2}\right)_t \qquad (8.5)$$

This is a partial differential equation involving the two independent variables x and t. Its numerical solution is a bit more complicated than those we have considered to this point.

8.1.2 Analytical solutions of Fick's second law in one dimension

Before proceeding to numerical solutions of Fick's second law, we write—without derivation—the analytical solution for a special but useful case, in which at time $t = 0$ the concentration is zero everywhere except at position x_0 where $c(x_0, t = 0) = c_0$.

It can be shown that the solution for this initial condition is

$$c(x,t) = \frac{c_0}{\sqrt{4\pi Dt}} e^{-(x-x_0)^2/4Dt} \tag{8.6}$$

This has the form of the familiar Gaussian function (see Section 3.2.1) with mean x_0 and standard deviation $\sigma = \sqrt{2Dt}$. As time proceeds, the initial concentration spike remains centered at x_0 but broadens proportionally to the *square root* of the time. It is this square-root behavior that distinguishes random diffusional motion from "purposive" or "driven" motion in which displacement is linearly proportional to the time.

Consider, for example, the concentration profile of an initial sharp spike of a small molecule such as sucrose as it diffuses in water at room temperature. According to the Stokes-Einstein equation for the diffusion of a sphere,

$$D = \frac{k_B T}{f} = \frac{k_B T}{6\pi \eta R} \tag{8.7}$$

where $k_B = 1.38 \times 10^{-16}$ erg/K is the Boltzmann constant in cgs units, $T = 293$ K is the Kelvin temperature, $\eta = 0.01$ poise is the viscosity of water at 20°C (= 293 K), and $R = 3 \times 10^{-8}$ cm is the hydrated molecular radius. A short calculation gives $D = 7.15 \times 10^{-6}$ cm^2/s, so that the rms distance σ the spike will have spread after one hour (3600 s) is 0.227 cm. We plot the concentration distribution at 10, 100, and 1000 seconds as follows. (Note that c throughout this section is the concentration as a function of position and time, not the combination or concatenation operator for forming a vector.)

```
> c = function(x,t,c0,x0,D){c0/sqrt(4*pi*D*t)*
  exp(-(x-x0)^2/(4*D*t))}
> x = seq(-1,1,0.05)
> c0 = 1; x0 = 0; D = 7.15e-6
> plot(x,c(x,10,c0,x0,D),type="l",xlab="x",
  ylab="c(x,t)")
> lines(x,c(x,100,c0,x0,D),lty=2)
> lines(x,c(x,1000,c0,x0,D),lty=3)
```

A common experimental situation, for example in establishing a sucrose density gradient, is to put a layer of pure water over a layer of sucrose solution. (The sucrose solution, being denser, is at the bottom.) If the sucrose concentration is initially uniform, it can be modeled as a superposition of spikes, each infinitely thin and displaced by an infinitesimal amount along the x-axis. We let the boundary between the solution and water be at $x = 0$, with the sucrose extending to negative x and the pure water to positive x:

$$c = c_0, \quad x < 0; \qquad c = 0, \quad x > 0; \qquad t = 0 \tag{8.8}$$

The solution of Fick's second law for this initial "step-function" condition can be obtained by integrating the solution eq. 8.6 over x to obtain [13, p. 14]

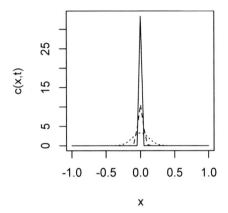

Fig. 8.1 Time evolution of a diffusing concentration spike

$$c(x,t) = \frac{1}{2}c_0 \operatorname{erfc} \frac{x}{2\sqrt{Dt}} \tag{8.9}$$

where erfc is the error function complement, defined as

$$\operatorname{erfc}(z) = 1 - \operatorname{erf}(z) = 1 - \frac{2}{\sqrt{\pi}} \int_0^z \exp(-w^2)dw = \frac{2}{\sqrt{\pi}} \int_z^\infty \exp(-w^2)dw \tag{8.10}$$

Conveniently, R has an easy way to calculate erf and erfc, albeit hidden away in R Help for `Normal`, the functions associated with the normal distribution in statistics:

```
erf = function(x)  2 * pnorm(x * sqrt(2)) - 1
erfc = function(x)  2 * pnorm(x * sqrt(2), lower = F)
```

The code to calculate and plot the time evolution of the concentration distribution for the step-function initial conditions is

```
> erfc = function(x)  2 * pnorm(x * sqrt(2), lower = F)
> c = function(x,t,c0,D){
c0/2*erfc(x/sqrt(4*D*t))
}
> x = seq(-1,1,0.05)
> c0 = 1;  D = 7.15e-5
> plot(x,c(x,10,c0,D),type="l",xlab="x",ylab="c(x,t)")
> lines(x,c(x,100,c0,D),lty=2)
> lines(x,c(x,1000,c0,D),lty=3)
```

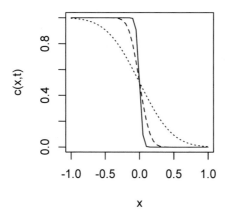

Fig. 8.2 Time evolution of an inital step-function concentration profile

8.1.3 Numerical solutions of Fick's second law in one dimension

Now let us show how to solve eq. 8.5 numerically to get the time- and position-dependence of the concentration of a diffusing species in a solution in which c varies in only one spatial dimension. To do so we use an elaboration of the finite difference method that we used in Euler's method. We divide "configuration space" into a two-dimensional grid with x running left-to-right in steps δx, and t running bottom-to-top in steps δt. We let $c_{i,j}$ be the concentration at position $i\delta x$ and time $j\delta t$. Expanding in Taylor's series and keeping only the leading terms, we have

$$c_{i,j+1} = c_{i,j} + \left(\frac{\partial c}{\partial t}\right)_{i,j} \delta t + \dots \tag{8.11}$$

and

$$c_{i+1,j} = c_{i,j} + \left(\frac{\partial c}{\partial x}\right)_{i,j} \delta x + \frac{1}{2}\left(\frac{\partial^2 c}{\partial x^2}\right)_{i,j} (\delta x)^2 + \dots \tag{8.12}$$

$$c_{i-1,j} = c_{i,j} - \left(\frac{\partial c}{\partial x}\right)_{i,j} \delta x + \frac{1}{2}\left(\frac{\partial^2 c}{\partial x^2}\right)_{i,j} (\delta x)^2 - \dots \tag{8.13}$$

$$\tag{8.14}$$

so the first-order finite-difference expression for the second derivative is

$$\left(\frac{\partial^2 c}{\partial x^2}\right)_{i,j} = \frac{c_{i+1,j} - 2c_{i,j} + c_{i-1,j}}{(\delta x)^2} \tag{8.15}$$

and the evolution of the concentration according to Fick's second law can be written in finite difference form as

$$c_{i,j+1} = c_{i,j} + \frac{D\delta t}{(\delta x)^2}\left(c_{i+1,j} - 2c_{i,j} + c_{i-1,j}\right) \tag{8.16}$$

Then the R code to calculate the numerical solution for the step function initial conditions is

```
> xmin = -1;xmax = 1;dx = 0.05;nx = (xmax-xmin)/dx + 1
> ix = seq(1:nx); x = seq(xmin,xmax,dx)
> tmin = 0;tmax = 1000;dt = 1;nt = (tmax-tmin)/dt + 1
> it = seq(1:nt); t = seq(tmin,tmax,dt)

> D = 7.15e-6; r = D*dt/dx^2; r
> c0 = 1

> c = matrix(rep(0,nx*nt), nrow = nx, ncol = nt)
> # Boundary condition at extreme left:
> c[1:floor(nx/2),1] = c0
> # plot(x,c[ix,1], type = "l")

> for (j in 1:(nt-1)) {
c[1,j+1] = c0
for (i in 2:(nx-1)) {
c[i,j+1] = c[i,j] + r*(c[i+1,j] - 2*c[i,j] + c[i-1,j])
}
}
> plot(x,c[ix,nt],type="l")
```

Note that we have had to add a boundary condition at the extreme left, if we want to keep the concentration there at its initial value. This corresponds to the left edge of the solution being far enough away from the step that diffusion does not affect it during the time of the experiment.

The numerical solution procedure we have just described is not stable if the time step is too large or the distance step is too small. As Crank [13, pp. 138, 143–144] discusses, if we define the dimensionless variables $X = x/l$ and $T = Dt/l^2$ where l is the length of the system, and let $r = \delta T/\delta X^2$ in eq. 8.16, then the solution is unstable if $r > 0.5$. The proper procedure in that case is to use the "Crank-Nicholson implicit method" [13, pp. 144–146].

8.1.4 Diffusion in spherical and cylindrical geometries

The real world, of course, is three-dimensional. Although one-dimensional diffusion may be a useful model for some experimental situations, other geometries may also be important [13, pp. 148–149]. The generalization of Fick's second law to three dimensions is

$$\left(\frac{\partial c(\mathbf{r},t)}{\partial t}\right)_{\mathbf{r}} = D\nabla_i^2 c(\mathbf{r},t) \tag{8.17}$$

which in Cartesian coordinates becomes

$$\left(\frac{\partial c}{\partial t}\right) = D\left[\left(\frac{\partial^2 c}{\partial x^2}\right) + \left(\frac{\partial^2 c}{\partial y^2}\right) + \left(\frac{\partial^2 c}{\partial z^2}\right)\right] \tag{8.18}$$

where we have omitted the subscripts for simplicity. Note that we have also assumed that D is independent of position or concentration.

For diffusion in an isotropic sphere of radius R (e.g., a spherical model of a cell), eq. 8.18 becomes

$$\frac{\partial c}{\partial t} = D\frac{1}{R^2}\frac{\partial}{\partial R}\left(R^2\frac{\partial c}{\partial R}\right) \tag{8.19}$$

which can be converted to the finite difference equation

$$c_{i,j+1} = c_{i,j} + \frac{D\delta t}{i(\delta R)^2}[(i+1)c_{i+1,j} - 2ic_{i,j} + (i-1)c_{i-1,j}], \quad i \neq 1 \tag{8.20}$$

$$= \frac{6}{(\delta R)^2}(c_{2,j} - c_{1,j}), \quad i = 1 \tag{8.21}$$

Another important case is diffusion in a two-dimensional solution or membrane with cylindrical symmetry. An example is the technique of fluorescence recovery after photobleaching, in which a laser bleaches the fluorophores in a circular area (typically about 50 μm in diameter) and the rate of recovery of fluorescence indicates the diffusion rate into the bleached area. If the area is cylindrically symmetric, Fick's second law becomes

$$\frac{\partial c}{\partial t} = D\frac{1}{R}\frac{\partial}{\partial R}\left(R\frac{\partial c}{\partial R}\right) \tag{8.22}$$

and the appropriate finite difference equation is

$$c_{i,j+1} = c_{i,j} + \frac{D\delta t}{2i(\delta R)^2}[(2i+1)c_{i+1,j} - 4ic_{i,j} + 2(i-1)c_{i-1,j}], \quad i \neq 1 \tag{8.23}$$

$$= \frac{4}{(\delta R)^2}(c_{2,j} - c_{1,j}), \quad i = 1 \tag{8.24}$$

Note that Crank's equations [13, pp. 148–149] start with $i = 0$, while ours start with $i = 1$, because matrices in R start indexing at 1.

8.2 Diffusion in a driving field: electrophoresis

In experimental techniques such as electrophoresis or sedimentation, an applied force acts in addition to the concentration gradient to drive material from a region of high potential to one of lower potential. In electrophoresis, the force is due to an applied voltage and is constant across the cell. In sedimentation, the centrifugal force increases linearly with distance from the origin (axis of rotation). The preceding model can be augmented to model electrophoresis by adding a term to the equation that describes the net flux of solute:

$$J = Vc - D\frac{\partial c}{\partial x} \tag{8.25}$$

where V is a constant proportional to the applied voltage and the electrophoretic mobility of the particle. Then the augmented form of Fick's second law is

$$\left(\frac{\partial c}{\partial t}\right)_x = -V\left(\frac{\partial c}{\partial x}\right)_t + D\left(\frac{\partial^2 c}{\partial x^2}\right)_t \tag{8.26}$$

with finite-difference representation

$$c_{i,j+1} = c_{i,j} - \frac{V}{\delta x}(c_{i+1} - c_i) + \frac{D\delta t}{(\delta x)^2}\left(c_{i+1,j} - 2c_{i,j} + c_{i-1,j}\right) \tag{8.27}$$

8.3 Countercurrent diffusion

Countercurrent diffusion involves two vessels or chambers in communication across a thin wall or membrane, through which material or heat can be exchanged by diffusion. The fluid in the two chambers flows in opposite directions, allowing more effective separation of material than could be achieved in a static flow system. This mechanism has evolved in the kidney for efficient transfer of solutes from high to low concentration. In a similar system called the *rete mirabile*, diving mammals such as whales and seals transfer heat between arterial and venous blood flowing in their flippers, thereby conserving heat.

This figure is a schematic diagram of countercurrent diffusion. Fluid enters the high-concentration (or heat the high-temperature) chamber H, and flows from left to right. Simultaneously, fluid enters the low-concentration chamber L and flows from right to left. H and L are in contact across a membrane (dotted line), and material flows across the membrane from compartments H_i to L_i according to Fick's first law. The division of H and L into compartments numbered from 1 to n is arbitrary for convenience, as are the assumptions that the volume of each compartment = 1 and the area of the membrane separating two compartments = 1. H_i and L_i designate both the compartments and the amounts or concentrations of material within them. Then the flow across the membrane from H_i to L_i in unit time is

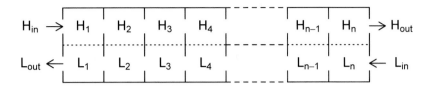

Fig. 8.3 Schematic diagram of countercurrent diffusion

$$J_i = D(H_i - L_i) \tag{8.28}$$

and the concentrations are updated after time Δt by

$$H_i(t + \Delta t) \leftarrow H_i(t) - J_i \Delta t \tag{8.29}$$
$$L_i(t + \Delta t) \leftarrow L_i(t) + J_i \Delta t \tag{8.30}$$

Within the H and L channels, fluid flows by so-called "plug flow", whereby each concentration is replaced by that of its neighbor to the left (H) or right (L):

$$H_{out} \leftarrow H_n, \cdots, H_{i+1} \leftarrow H_i, \cdots, H_1 \leftarrow H_{in} \tag{8.31}$$
$$L_{out} \leftarrow L_1, \cdots, L_{i-1} \leftarrow L_i, \cdots, L_n \leftarrow L_{in} \tag{8.32}$$

In Problem 3, you are asked to write an R program to simulate countercurrent diffusion and plot H_i and L_i after some time has passed.

8.4 Diffusion as a random process: Brownian motion

At the microscopic level, diffusion is a random thermal process as the particles of interest are buffeted by solvent molecules and undergo Brownian motion. The particles appear to undergo a random walk. The higher the temperature, the more vigorously the solvent molecules collide, hence the faster diffusion. Macroscopically, Fick's first law says that particles diffuse from regions of high concentration to regions of lower concentration in response to a concentration (or chemical potential) gradient. Microscopically, each particle moves with equal probability to the left or to the right; but if there are more particles on the left, there will be a net flow to the right.

Since each particle moves with equal probability to the left or right (up or down, in and out in two or three dimensions), its mean displacement $<x>$ as a function of time is zero. However, the mean-square displacement $<x^2>$ is always positive. Integration of eq. 8.6 gives for the one-dimensional case

$$< x^2 >= \int_{-\infty}^{\infty} x^2 c(x) dx = 2Dt \qquad (8.33)$$

The corresponding result is $< R^2 >= 4Dt$ in two dimensions, and $< R^2 >= 6Dt$ in three dimensions.

Diffusion can be simulated as a random walk. We consider a two-dimensional random walk, since the particle path can be readily plotted. As one possible model, we restrict the steps along the x and y axes to ± 1, so that the particle does a random walk on a square lattice. We start the particle at the origin, and track its path over n steps.

```
> n = 100
> x = 0; y = 0 # Particle initially at origin
> xsteps = sample(c(-1,1),size=n,replace=TRUE,prob=c(0.5,0.5))
> ysteps = sample(c(-1,1),size=n,replace=TRUE,prob=c(0.5,0.5))
> for (i in 1:n) {
x[i+1] = x[i] + xsteps[i]
y[i+1] = y[i] + ysteps[i]
}
> plot(x,y,type="l")
> rms.dist = sqrt((x[n+1]-x[1])^2 + (y[n+1]-y[1])^2)
> rms.expected = sqrt(2*n)
```

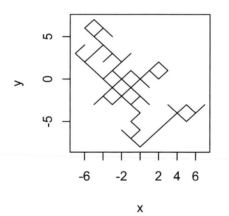

Fig. 8.4 Example of a 100-step random walk on a square lattice

For this particular run, the root-mean-square end-to-end distance was 7.2, compared with the expected value of $\sqrt{2n} = 14.14$ for a random walk with step length $\sqrt{2}$ and $n = 100$ steps. Other runs would generate different random numbers and therefore different rms distances.

There are other ways to model diffusion. One is to pick a single random number θ between 0 and 2π, and let the changes in x and y be $\delta x = b \sin(\theta)$ and $\delta y = b \cos(\theta)$

where b is the step length. Another, more consistent with the random nature of the diffusion process, would be to let the step length be drawn from the normal distribution and the step direction from a uniform distribution. These variants are considered in Problem 4.

8.5 Compartmental models in physiology and pharmacokinetics

The finite difference method we have used to calculate diffusion and electrophoresis computes the flow of material into and out of small volumes of space. We can think of those volumes as "compartments", and view flow processes as the transport of material between compartments.

In fact, compartmental models are widely used in some areas of physiology and medicine, especially pharmacokinetics, to analyze the flow of drugs, tracer compounds, and metabolites between biologically relevant "compartments" such as blood, liver, kidney, and urine. The materials of interest are introduced into one compartment, are stored or metabolized there or transported to other compartments. Whatever has not been stored or reacted is eventually excreted in one form or another. What is commonly measured is the amount continuing to circulate in the blood or excreted in the urine.

The approximations typically used in compartmental models are severe. The compartments are generally treated as homogeneous, well-stirred, and of constant volume, so that organs as spatially extended and complex as blood or liver are treated as uniform "boxes". (There is a class of models involving the lungs and heart in which the volume changes in response to pressure; we will not consider them here.) Such approximations are often necessary both because of our lack of knowledge of true structures and concentration distributions, and because more realistic models would involve a multitude of parameters, most of them inaccessible to experimental measurement.

In the following sections we examine some elementary compartmental models that are widely used in the literature. Our treatment draws substantially on that of Keen and Spain [39, Ch. 15]. More comprehensive treatments can be found in [9] and [42]. R has a number of packages developed to deal with pharmacokinetic data and modeling. These are surveyed in the "CRAN Task View: Analysis of Pharmacokinetic Data" available through the Task Views link on the CRAN web site home page. The packages PKfit and PKtools are particularly pertinent to the discussion in this section, for those who wish to delve deeper into the topic.

First some terminology: A one-compartment model is a single, well-mixed container of volume V containing amount Q of the compound of interest, whose concentration is $c = Q/V$. If V is in mL and Q in mg, then the units of c are mg/mL. If there is a constant feed into the compartment, it is characterized by a rate constant k_{in} with units of mg/time. The same units apply to the excretion rate constant k_{out}. Alternatively, there may be discrete inputs (doses) at specified times, with units of mg. In either case, it is assumed that the added compound is instantly well-mixed throughout the compartment. Of course, amounts may be measured in other mass units or in molar units. In two- or multi-compartnent models, inputs to and outputs from compartment i are specified with subscript i; and the rate constant for transfer from compartment i to compartment j is k_{ij}.

8.5.1 Periodic dose administration to a single compartment

As an example of how a single-compartment model can provide insight into the time course of drug availability in a target body system, consider the daily administration of a drug S intended to affect blood pressure. The "compartment" in this case consists of those organs that control blood pressure, e.g., kidneys, blood vessels, and heart. A pill is taken each morning, starting at $t = 0$ and repeating at each $t =$ an integer multiple of 24 hours. It immediately increases the quantity of S in the compartment by an amount ΔS. The drug is lost by metabolism and excretion in a rate process that is related to the amount in the compartment but is assumed to be saturable:

$$\frac{dS}{dt} = -k_{out}\frac{S}{K+S} \tag{8.34}$$

Problem 6 asks you to compute the time-dependence of S in the compartment. You should observe approach to a fluctuating plateau over several weeks.

8.5.2 Liver function—A two-compartment model

An example of a model involving transfer between two compartments—blood and liver—is the test for liver function involving the dye tracer bromosulfophthalein (BSP). The dye is injected into the bloodstream and should be cleared from the circulation through conjugation to sulfhydryl groups by a normally functioning liver. The amount of dye remaining in the blood plasma at various times after injection is measured spectrophotometrically.

In the model, a single dose D of BSP is injected into the blood (compartment 1) at time $t = 0$. The dye is transferred from blood to liver (compartment 2) with rate constant k_{12}. Once in the liver, the BSP is either conjugated and excreted (rate constant $k_{2,out}$) or exchanged back to the blood plasma (rate constant k_{21}). For each process, it is assumed that the rate is proportional to the amount of BSP in the compartment; this is called donor-controlled transport. The kinetic rate equations are then

$$\frac{dc_1}{dt} = -k_{12}c_1 + k_{21}c_2$$
$$\frac{dc_2}{dt} = k_{12}c_1 - k_{21}c_2 - k_{2out}c_2 \tag{8.35}$$

with initial conditions $c_1(0) = D$ and $c_2(0) = 0$. In Problem 7, you are asked to use these equations to calculate the concentration of BSP remaining in the blood plasma as a function of time after injection.

8.5.3 Multi-compartment model of liver function

More detail about liver function was obtained by Molino and Milanese [46, 45], as summarized by Keen and Spain [39]. They studied BSP processing in human subjects whose gallbladders had been removed previously. They were able to measure unconjugated and conjugated BSP in the liver and in the bile tracts inside the liver, as well as in the blood plasma, and developed a six-compartment model:

Number	Compartment
1	Blood plasma with unconjugated BSP
2	Liver with unconjugated BSP
3	Bile tracts with unconjugated BSP
4	Blood plasma with conjugated BSP
5	Liver with conjugated BSP
6	Bile tracts with conjugated BSP

Conversion from unconjugated to conjugated bile occurs in the liver. Blood and liver exchange both unconjugated and conjugated BSP reversibly, while transfer from liver to bile tracts is irreversible. Excretion occurs from the bile tracts, compartments 3 and 6. The rate equations for the changes in amounts of BSP in the six compartments are

$$\frac{dc_1}{dt} = -k_{12}c_1 + k_{21}c_2 \tag{8.36}$$

$$\frac{dc_2}{dt} = k_{12}c_1 - (k_{21} + k_{23} + k_{25})c_2 \tag{8.37}$$

$$\frac{dc_3}{dt} = k_{23}c_2 - k_{3out}c_3 \tag{8.38}$$

$$\frac{dc_4}{dt} = -k_{45}c_4 + k_{54}c_5 \tag{8.39}$$

$$\frac{dc_5}{dt} = k_{25}c_2 + k_{45}c_4 - (k_{54} + k_{56})c_5 \tag{8.40}$$

$$\frac{dc_6}{dt} = k_{56}c_5 - k_{6out}c_6 \tag{8.41}$$

In Problem 8 you are asked to simulate this model for given values of the rate constants.

8.5.4 Oscillations in calcium metabolism

Studying calcium-deficient rats with ^{45}Ca as tracer, Staub and coworkers [58] found that calcium levels in blood and bone fluctuate much more than in normal rats, with an approximately 24-hour period. They found that a model consisting of at least five compartments—blood plasma and four bone compartments (a–d)—was needed to explain the data. In that model, blood plasma reversibly exchanges Ca with bone compartments a and d, a exchanges with b in a nonlinear fashion that provides the oscillatory behavior, and b exchanges with c. If we denote the ^{45}Ca concentration in blood plasma as c_1, and the ^{45}Ca concentrations in bone compartments a–d as c_2–c_5, the rate equations characterizing the model are

$$\frac{dc_1}{dt} = -(k_{12}+k_{15})c_1 + k_{21}c_2 + k_{51}c_5 \tag{8.42}$$

$$\frac{dc_2}{dt} = k_{12}c_1 - k_{21}c_2 - k_{23}c_2c_3^2 + k_{32}c_3 \tag{8.43}$$

$$\frac{dc_3}{dt} = k_{23}c_2c_3^2 - (k_{32}+k_{34})c_3 + k_{43}c_4 \tag{8.44}$$

$$\frac{dc_4}{dt} = k_{34}c_3 - k_{43}c_4 \tag{8.45}$$

$$\frac{dc_5}{dt} = k_{15}c_1 - k_{51}c_5 \tag{8.46}$$

There is no flow to or from the external environment in this model. In Problem 9 you are asked to demonstrate the oscillations in this model for given values of the rate constants and initial conditions.

8.6 Problems

1. Plot the analytical and numerical solutions to the step-function diffusion conditions on the same graph, overlaying a line for the analytical solution on points for the numerical solution. Use tmax=1000.
2. Redo the previous problem with the flux equation including the electrophoresis term, with $V = 10\ \mu$m/s.
3. Write an R program for countercurrent diffusion using 20 pairs of H and L compartments ($n = 20$), with $D = 0.1$, $H_{in} = 100$, and $L_{in} = 0$. Use initial conditions $H_i = 0$ and $L_i = 0$, $i = 1 \ldots n$. Let $\Delta t = 1$, run the simulation for 50 time intervals, and then plot H_i and L_i vs i.
4. Use the other ways suggested in the text to model 2-D diffusion for 100 steps. (a) Pick random numbers θ between 0 and 2π, and let the changes in x and y be $\delta x = b\sin(\theta)$ and $\delta y = b\cos(\theta)$ where b is the step length. (b) Draw the step length from a normal distribution with mean 1 and sd 0.5, and the step angle from a uniform distribution between 0 and 2π.

5. Modify the 2-D Brownian motion diffusion simulation to plot the distribution of rms displacements of 100 particles, each of which has taken 100 steps starting at the origin. What is the mean displacement? Fit the distribution to a normal distribution.

6. Calculate and plot the compartmental amount of a blood pressure drug S that is administered every 24 hours, starting at $t = 0$, at a dose $D = 1$ mg, and is eliminated according to eq. 8.34 with $k_{out} = 0.1$ mg/day and $K = 10$ mg. Use two-step Euler integration with a time increment of 0.1 day, and let the calculation run for 1 month (720 hours). To check whether exactly an integer multiple of 24 hours has passed, use the *modulo* function %/% which yields the integer quotient of two numbers.

7. Use eqs. 8.35 to calculate and plot the concentration (mg/liter) of BSP from the blood plasma over a period of 2 hours. Let the initial injection D of BSP be 350 mg, and the plasma volume be 3.5 liters. The rate constants are $k_{12} = 0.14$ min^{-1}, $k_{21} = 0.0038$ min^{-1}, and $k_{2out} = 0.032$ min^{-1}. Also plot the log of the BSP concentration vs. time; the curved semilog plot indicates the presence of two (or more) rate processes.

8. Develop an R simulation of the multi-compartment model of liver function, eqs. 8.36–8.41. Use the values of the transfer rate constants (all units min^{-1}) determined experimentally by Molino and Milanese [46]:
$k_{12} = 0.1130, k_{21} = 0.0300, k_{23} = 0.0043, k_{3out} = 0.0480$
$k_{45} = 0.0078, k_{54} = 0.9900, k_{56} = 0.6280, k_{6out} = 0.0580$
$k_{25} = 0.0090$
Give a single intravenous injection of 300 mg BSP at $t = 0$. BSP, either unconjugated or conjugated, does not appear in the bile until 20 minutes have passed. Simulate this time delay by setting k_{3out} and $k_{6out} = 0$ until 20 minutes have passed. Plot the mg BSP in each compartment and in the environment over a period of 3 hours (180 minutes).

9. Use eqs. 8.42–8.46 to model the flow of calcium in Ca-deficient rats, using the rate constants (units min^{-1} except for k_{23}) determined by Staub et al [58]
$k_{21} = 8.506 \times 10^{-2}, k_{12} = 6.534 \times 10^{-1}, k_{51} = 1.951 \times 10^{-2}$
$k_{32} = 6.019 \times 10^{-1}, k_{23} = 7.533 \times 10^{-3}, k_{15} = .2922 \times 10^{-1}$
$k_{43} = 7.190 \times 10^{-3}, k_{34} = 2.816 \times 10^{-1}$
and initial amounts (mg) in each compartment
$Q_1 = 1.558, Q_2 = 11.792, Q_3 = 6.079, Q_4 = 224.537, Q_5 = 25.534.$
Plot the amount of Ca in each compartment over 4 days (96 hours). You will need a small δt to handle the nonlinearities in this system. Start with $\delta t = 0.01$ hr and increase it until the calculation fails to converge.

Chapter 9
Regulation and Control of Metabolism

Metabolism involves not just single biochemical reactions, but coordinated networks of reactions. These networks are usually remarkably well-regulated, keeping close to a set-point, a steady-state or dynamic equilibrium in most healthy organisms. Substantial deviation from that set-point may betoken disease or some other extraordinary circumstance. On the other hand, biotechnologists may want to manipulate an organism to overproduce a desirable product, controlling its metabolism to deviate from the normal set-point. Groen and Westerhoff [30] have listed four questions that theories of metabolic control should answer:

1. Which steps exert significant control on pathway flux?
2. What is the control by the enzymes on metabolite concentrations?
3. What is the underlying mechanism for the control the enzymes exert?
4. Which regulatory mechanism is most important for pathway control under physiological conditions?

In this chapter we examine these issues of regulation and control by simulating the behavior of networks of enzymatic reactions. This is a complicated area, so our treatment will be able only to scratch the surface. For more information, the books by Beard and Qian [6], Fell [25], Voit [66], and Torres and Voit [63] will be useful.

9.1 Successive enzyme reactions

A key point to realize about sequences or networks of enzymatic reactions is that they are generally operating under quite different conditions than a single reaction in a kinetic study. Typically the goal of a kinetic study of an enzyme is to determine its maximum velocity V_{max} and Michaelis constant K_m. This is most often done by measuring the substrate-dependence of the initial velocity, before product has had time to build to a significant concentration. In contrast, metabolic networks almost by definition have products as well as substrates in the mixture, since the product of one reaction is the substrate of the next. The reactions may in many cases be

V. Bloomfield, *Computer Simulation and Data Analysis in Molecular Biology and Biophysics,*   175
Biological and Medical Physics, Biomedical Engineering, DOI: 10.1007/978-1-4419-0083-8_9,
© Springer Science + Business Media, LLC 2009

close to equilibrium, or in any case be in a steady state where the reaction flux (net rate of product production) is essentially the same from one reaction to the next. Thus the dependence of the flux on enzyme or substrate concentration will generally be very different from that of an isolated enzyme. Product inhibition is likely to be important, and the enzyme may be operating with substrate near or above K_m, perhaps close to saturation.

9.1.1 One-substrate, one-product reaction

Consider, for example, a reversible 1-substrate, 1-product enzyme-catalyzed reaction in which substrate is fed in and product taken out at a constant flux J.

$$J \to S \overset{E}{\rightleftharpoons} P \to J \tag{9.1}$$

We define the net velocity v of the enzyme-catalyzed reaction as a function of the substrate S and product P concentrations and the kinetic parameters (maximum velocities and Michaelis constants in both forward and reverse directions), give those parameters and the flux J arbitrary values, and set initial values for the substrate and product concentrations. We then use the Euler method to solve the time-dependence of the concentrations over a sufficient length of time that a steady state is reached, plot the results, and compare the steady-state and equilibrium concentrations.

```
> v = function(S,P,parm) {
  Vmf = parm[1]; Vmr = parm[2]; Kmf = parm[3]; Kmr = parm[4]
  (Vmf*S/Kmf - Vmr*P/Kmr)/(1+S/Kmf+P/Kmr)
  }

> parm = c(Vmf = 0.5, Vmr = .05, Kmf = 1, Kmr = 1)

> J = Vmf/10

> S = Kmf; P = Kmr # Initial values

> tmin = 0; tmax = 50; dt = 0.1; n = (tmax-tmin)/dt + 1
> t = seq(tmin,tmax,dt)

> for (i in 2:n) {
  dPdt = v(S[i-1],P[i-1],parm) # Net rate of enzyme reaction
  P[i] = P[i-1]+dPdt*dt - J*dt # Product removed by flux J
  S[i] = S[i-1] - dPdt*dt + J*dt # Substrate fed in by J
  }

> plot(t,S,type="l", ylim = c(0,max(c(S,P))),ylab="Conc")
> lines(t,P,lty=2)
> legend(0.6*tmax,0.8*max(c(S,P)),legend=c("S","P"),
    lty=c(1,2),bty="n")

> c(S[n], P[n]) # Steady state concentrations
```

```
[1] 0.4546 1.5454
> P[n]/S[n]   # Steady state ratio
[1] 3.399

> Keq = Vmf*Kmr/(Vmr*Kmf) # Briggs-Haldane relation
> Keq  # Equilibrium ratio
[1] 10
```

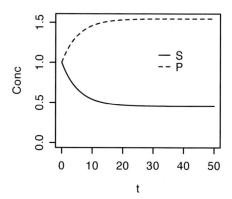

Fig. 9.1 Time evolution to the steady state of mechanism 9.1

Under these conditions, the steady-state ratio is not equal to the equilibrium ratio. The steady-state ratio can be calculated, without simulation of the reaction kinetics, by noting that the rate of disappearance of S plus its feed flux J must equal zero, and the amount of S plus P must remain constant. With the parameters above and a nonlinear minimization routine like nlm, we can solve for the steady-state values of S and P that satisfy these conditions.

```
> fun = function(x,parm,J){
S = x[1];  P = x[2]
sq1 = (-v(S,P,parm)+J)^2 # Net change of S
sq2 = (S + P - (Kmf + Kmr))^2 # Conservation of S + P
return(sq1+sq2)
}

> S.ss = nlm(fun,c(1,1),parm,J)$estimate[1]
> P.ss = nlm(fun,c(1,1),parm,J)$estimate[2]
> c(S.ss, P.ss)
[1] 0.4546 1.5454
> P.ss/S.ss
[1] 3.4
```

The results agree with those from the kinetic simulation.

The disequilibrium ratio ρ is the ratio of steady-state concentrations to equilibrium concentrations,

$$\rho = \frac{V_{mr}K_{mf}}{V_{mf}K_{mr}}\frac{P_{ss}}{S_{ss}}. \tag{9.2}$$

For this example, $\rho = 3.4/10 = 0.34$ compared to a value of 1 if the reaction were at equilibrium.

9.1.2 Successive one-substrate, one-product reactions

We extend this treatment to a series of two successive reactions

$$J \to S \underset{}{\overset{E_1}{\rightleftharpoons}} P \underset{}{\overset{E_2}{\rightleftharpoons}} Q \to J \tag{9.3}$$

```
> v = function(S,P,parm) {
Vmf = parm[1]; Vmr = parm[2]; Kmf = parm[3]; Kmr = parm[4]
(Vmf*S/Kmf - Vmr*P/Kmr)/(1+S/Kmf+P/Kmr)
}

> # Kinetic parameters for reactions 1 and 2
    > parms1 = c(Vmf1 = 5, Vmr1 = .5, Kmf1 = 1, Kmr1 = 1)
> parms2 = c(Vmf2 = 3, Vmr2 = 1, Kmf2 = 2, Kmr2 = .5)
> J = Vmf1/100

> S = Kmf1; P = Kmr1; Q = Kmr2 # Initial values

> tmin = 0; tmax = 50; dt = 0.1; n = (tmax-tmin)/dt + 1
> t = seq(tmin,tmax,dt)

> for (i in 2:n) {
v1 = v(S[i-1],P[i-1], parms1)
v2 = v(P[i-1],Q[i-1],parms2)
dSdt = -v1
dPdt = v1-v2
dQdt = v2
S[i] = S[i-1] + dSdt*dt + J*dt
P[i] = P[i-1] + dPdt*dt
Q[i] = Q[i-1] + dQdt*dt - J*dt
}

> plot(t,S,type="l", ylim = c(0,max(c(S,P,Q))),ylab="Conc")
> lines(t,P,lty=2)
> lines(t,Q,lty=3)

> legend(0.6*tmax,0.8*max(c(S,P,Q)),legend=c("S","P","Q"),
lty=c(1,2,3),bty="n")

> c(S[n], P[n], Q[n]) # Steady-state concentrations
[1] 0.1641 1.3861 0.9498
```

```
> c(P[n]/S[n], Q[n]/P[n])   # Steady-state ratios
[1] 8.4461 0.6852

> Keq1 = Vmf1*Kmr1/(Vmr1*Kmf1)
> Keq2 = Vmf2*Kmr2/(Vmr2*Kmf2)
> c(Keq1, Keq2)   # Equilibrium ratios
[1] 10.00   0.75
```

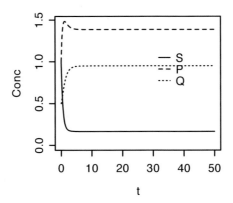

Fig. 9.2 Time evolution to the steady state of mechanism 9.3

Again, we check against a direct calculation of the steady-state ratios, and find excellent agreement.

```
> fun = function(x, parms1,parms2,J){
S = x[1]; P = x[2]; Q = x[3]
sq1 = (-v(S,P,parms1)+J)^2 # Net change of S
sq2 = (v(P,Q,parms2)-J)^2 # Net change of Q
# Conservation of material:
sq3 = (S + P + Q - (Kmf1 + Kmr1 + Kmr2))^2
return(sq1+sq2+sq3)
}

> nlmest = nlm(fun,c(1,1,1),parms1,parms2,J)$estimate
> S.ss = nlmest[1]
> P.ss = nlmest[2]
> Q.ss = nlmest[3]
> c(S.ss,P.ss,Q.ss) # Steady-state concentrations
[1] 0.1641 1.3861 0.9498
> c(P.ss/S.ss, Q.ss/P.ss) # Steady-state ratios
[1] 8.4461 0.6852
```

The disequilibrium ratios are

```
# Disequilibrium ratios
> rho.1 = (P.ss/S.ss)/Keq1; rho.1
[1] 0.8446
> rho.2 = (Q.ss/P.ss)/Keq2; rho.2
[1] 0.9136
```

9.1.3 Steady-state flux calculation

Instead of specifying the flux, we can set the concentrations of the first substrate (X) and last product (Y) to some constant values, and calculate the steady-state flux as well as the concentrations. We expand the sequence to four successive one-substrate, one-product reactions, and generate the kinetic parameters (four for each reaction) with random numbers. The set.seed function enables us to replicate the random numbers. We also set the values of X and Y.

```
> set.seed(1)
> parms1 = round(runif(4),2)
> parms2 = round(runif(4),2)
> parms3 = round(runif(4),2)
> parms4 = round(runif(4),2)

> parms1; parms2; parms3; parms4

[1] 0.27 0.37 0.57 0.91
[1] 0.20 0.90 0.94 0.66
[1] 0.63 0.06 0.21 0.18
[1] 0.69 0.38 0.77 0.50

> X = 1; Y = 0.5
```

We proceed as in the prior examples to calculate the steady-state velocities and concentrations, which then enable calculation of the fluxes.

```
> v = function(S,P,parm) {
Vmf=parm[1]; Vmr=parm[2]; Kmf=parm[3]; Kmr=parm[4]
(Vmf*S/Kmf - Vmr*P/Kmr)/(1+S/Kmf+P/Kmr)
}

> fun = function(x,parms1,parms2,parms3,parms4,X,Y){
S1 = x[1]; S2 = x[2]; S3 = x[3]
v1 = v(X,S1,parms1)
v2 = v(S1,S2,parms2)
v3 = v(S2,S3,parms3)
```

```
v4 = v(S3,Y,parms4)
return((v1-v2)^2 + (v2-v3)^2 + (v3-v4)^2)
}
> nlmest = nlm(fun,c(1,1,1),parms1,parms2,parms3,
  parms4,X,Y)$estimate
> S1.ss = nlmest[1]
> S2.ss = nlmest[2]
> S3.ss = nlmest[3]
> c(S1.ss,S2.ss,S3.ss) # Steady-state concentrations
[1] 0.90334 0.09705 0.50835

# Steady-state ratios
> c(S1.ss/X,S2.ss/S1.ss, S3.ss/S2.ss, Y/S3.ss)
[1] 0.9033 0.1074 5.2379 0.9836

> # Fluxes through each enzyme-catalyzed step
> v(X,S1.ss,parms1)
[1] 0.02839
> v(S1.ss,S2.ss,parms2)
[1] 0.02839
> v(S2.ss,S3.ss,parms3)
[1] 0.02839
> v(S3.ss,Y,parms4)
[1] 0.02839
```

The fluxes all have the same steady-state value, as expected. Finally, we calculate the equilibrium ratios of concentrations.

```
> Vmf1=parms1[1];Vmr1=parms1[2];Kmf1=parms1[3];Kmr1=parms1[4]
> Vmf2=parms2[1];Vmr2=parms2[2];Kmf2=parms2[3];Kmr2=parms2[4]
> Vmf3=parms3[1];Vmr3=parms3[2];Kmf3=parms3[3];Kmr3=parms3[4]
> Vmf4=parms4[1];Vmr4=parms4[2];Kmf4=parms4[3];Kmr4=parms4[4]

> # Equilibrium ratios
> c(Vmf1*Kmr1/Vmr1/Kmf1, Vmf2*Kmr2/Vmr2/Kmf2,
  Vmf3*Kmr3/Vmr3/Kmf3, Vmf4*Kmr4/Vmr4/Kmf4)
[1] 1.1650 0.1560 9.0000 1.1791
```

9.2 Metabolic control analysis

Metabolic control analysis, founded on the work of Kacser and Burns [37] and Heinrich and Rapoport [32], is one of the major theoretical approaches to understanding the control of metabolic networks. We shall here examine just two of its key concepts: flux control coefficients and elasticities. The books by Heinrich and Schuster [33] and Fell [25, Ch. 5] have extensive treatments.

9.2.1 Flux control coefficients

A question often asked by biochemists is "What is the rate-limiting step in this pathway?" Answers given are often "The slowest enzyme, that with the smallest V_{max}" or "The one operating furthest from equilibrium." In fact, there is no general answer; the flux through a pathway is regulated by all the enzymes in the pathway, each contributing some proportion of the final result. Metabolic control analysis formalizes this conclusion by defining flux control coefficients

$$C_x^{J_y} = \frac{\partial J_y / J_y}{\partial E_x / E_x} = \frac{\partial \ln J_y}{\partial \ln E_x} \tag{9.4}$$

that is, the relative change in flux of reaction y produced by a relative change in the activity of enzyme E_x. In the steady state, all the Js in a given pathway are the same, so we will drop the y subscript. Also, we will assume that enzyme activities are proportional to their maximum velocities, so we replace relative changes in E with relative changes in V_{max}. (Changes in enzyme activity may also result from inhibition, oligomerization, etc. These possibilities are neglected here.)

Experimentally, one might measure the flux control coefficient attributable to an enzyme in a pathway by slightly increasing the amount of that enzyme and measuring the resulting increase in flux. We can simulate that experiment by increasing the values of V_{max} for the forward and reverse reactions catalyzed by the enzyme by a small amount (1% in the example below) and calculating the increase in steady-state flux. We start with the parameters in the system developed in the previous section, increase Vmf1 and Vmr1 by 1%, solve for the new steady-state concentrations and fluxes.

```
# Start with parameters for reaction 1
> parms1.a = parms1
# Increase Vmr, Vmf
> parms1.a[1]=1.01*parms1[1];parms1.a[2]=1.01*parms1[2]
> # Solve for new steady-state concentrations
> nlmest.a = nlm(fun,c(1,1,1),parms1.a,parms2,parms3,
  parms4,X,Y)$estimate
> S1.ss.a = nlmest.a[1]
> S2.ss.a = nlmest.a[2]
> S3.ss.a = nlmest.a[3]

    # Steady-state concentrations
> c(S1.ss.a,S2.ss.a,S3.ss.a)
[1] 0.90516 0.09719 0.50855

> # Solve for new fluxes
> v(X,S1.ss.a,parms1.a)
[1] 0.02846
> v(S1.ss.a,S2.ss.a,parms2)
```

```
[1] 0.02846
> v(S2.ss.a,S3.ss.a,parms3)
[1] 0.02846
> v(S3.ss.a,Y,parms4)
[1] 0.02846
```

We then calculate the relative change in J and divide it by the relative change in E (or V_{max}) to get the control coefficient C_1^J:

```
# Change in flux
> v(X,S1.ss.a,parms1.a)-v(X,S1.ss,parms1)
[1] 6.92e-05
> # Relative change in v1 or J
> dlnJ = (v(X,S1.ss.a,parms1.a) -
  v(X,S1.ss,parms1))/v(X,S1.ss,parms1); dlnJ
[1] 0.002437
> # Relative change in Vmf1 or E
> dlnE1 = (parms1.a[1]-parms1[1])/parms1[1]; dlnE1
[1] 0.01
> dlnJ/dlnE1
[1] 0.2437
```

Repeating this calculation for reactions 2, 3, and 4 gives dlnJ/dlnE2 = 0.3276, 0.3259, and 0.1188, respectively. The sum of all four flux control coefficients should equal 1.0 according to the summation theorem of metabolic control analysis [25, p. 110]. We find $0.2437 + 0.3276 + 0.3259 + 0.1188 = 1.016$, which is pretty close. The 1.6% discrepancy is due to the relatively large 1% change in enzyme concentration. Note that the relative size of the flux control coefficients bears no obvious relation to the values of the kinetic parameters. The coefficients depend on all the parameters and on the values of the steady-state concentrations of metabolites.

9.2.2 Elasticities

Flux control coefficients pertain to the entire reaction sequence. Elasticities connect the coefficients to the individual enzyme properties. The elasticity of reaction i with respect to substrate S is defined as

$$\varepsilon_S^i = \frac{\partial v_i/v_i}{\partial S/S} = \frac{\partial \ln|v_i|}{\partial \ln S} \tag{9.5}$$

where the absolute value of the logarithm of the velocity is required because the velocity might be negative. Elasticities, like flux control coefficients, are defined as *relative* changes in velocities or fluxes with respect to *relative* changes in a parameter.

For a reversible, one-substrate, one-product reaction of the sort we have been considering, it can be shown [25, p. 115] that

$$\varepsilon_S^v = \frac{1}{1-\rho} - \frac{S/K_{mf}}{1+S/K_{mf}+P/K_{mr}} = \frac{1}{1-\rho} - \frac{v_f}{V_{mf}} \tag{9.6}$$

where ρ is the disequilibrium ratio (eq. 9.2) and v_f is the total forward rate (not the net forward rate) of the reaction. Likewise, the product elasticity is given by

$$\varepsilon_S^v = \frac{-\rho}{1-\rho} - \frac{P/K_{mr}}{1+S/K_{mf}+P/K_{mr}} = \frac{-\rho}{1-\rho} - \frac{v_r}{V_{mf}} \tag{9.7}$$

With the same parameters as in the previous section, we calculate the elasticity of the forward direction of reaction 3 ($S_2 \rightarrow S_3$) with S_2 as substrate.

```
> Keq.3 = Vmf3*Kmr3/Vmr3/Kmf3; Keq.3 # Equilibrium constant
[1]9.0000
> rho.3 = (S3.ss/S2.ss)/Keq.3; rho.3 # Disequilibrium ratio
[1] 0.5820
> # Elasticity
> eps.S2.f = 1/(1-rho.3) + S2.ss/Kmf3/(1+S2.ss/Kmf3+S3.ss/Kmr3)
> eps.S2.f
[1] 2.5
```

The flux control coefficients and elasticities are connected through the connectivity theorem [37]:

$$\sum_{i=1}^{n} C_i^J \varepsilon_S^i = 0 \tag{9.8}$$

where the sum is over all of the enzymatic reactions i in which metabolite S is involved. For example, S_2 is a product in reaction 2 and a substrate in reaction 3. In problem 4 at the end of this chapter, you will be asked to check the validity of eq. 9.8 for this example.

9.3 Biochemical systems theory

Another approach to the integrated kinetic behavior of metabolic networks is provided by biochemical systems theory [54, 66, 63]. In this theory, the rate equations for the time-dependence of the metabolite concentrations are written as sums and products of powers of the concentrations. The powers may not be simple positive integers as is normal for reaction mechanisms. The power with which a substrate enters an enzyme-catalyzed reaction may be near 1 if its concentration is much less than K_m, about 0.5 if it is near K_m, and near 0 if it is close to saturating. Negative powers denote inhibition.

9.3.1 Linear pathway with feedback inhibition

A simple example is the linear pathway with feedback inhibition.

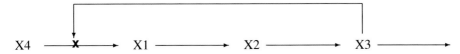

Rate equations with specific values of the rate constants and powers [63, p. 58] are

$$\frac{dX_1}{dt} = 32X_4^2 X_3^{-1} - 2X_1^{0.5}$$

$$\frac{dX_2}{dt} = 2X_1^{0.5} - 2X_2^{0.5} \tag{9.9}$$

$$\frac{dX_3}{dt} = 2X_2^{0.5} - 2 * X_3$$

X_4 is held constant at 0.25, except when it changes to 0.75 at $t = 1$. Initial values of X_1, X_2, and X_3 are all equal to unity. The R code to implement this mechanism is

```
> # Linear pathway with feedback inhibition
> # Torres and Voit, p. 58

> X10 = 1; X20 = 1; X30 = 1; X40 = 0.25 # Initial concs

> tmin = 0; tmax = 10; dt = 0.05; n = (tmax-tmin)/dt + 1
> t = seq(tmin,tmax,dt)

> X4.off = X40; X4.on = X40 + 0.5

> # Times to change X4 concentration
> t.on = 1; i.on = (t.on - tmin)/dt + 1
> t.off = 1.05; i.off = (t.off - tmin)/dt + 1

#  Establish variables
> X1 = X10; X2 = X20; X3 = X30; X4 = X40

> for (i in 2:n) {
     dX1 = (32*X4[i-1]^2*X3[i-1]^(-1) - 2*X1[i-1]^0.5)*dt
     dX2 = (2*X1[i-1]^0.5 - 2*X2[i-1]^0.5)*dt
dX3 = (2*X2[i-1]^0.5 - 2*X3[i-1])*dt
X1[i] = X1[i-1] + dX1
X2[i] = X2[i-1] + dX2
X3[i] = X3[i-1] + dX3
if (i >= i.on & i < i.off) X4[i] = X4.on else X4[i] = X4.off
}

> plot (t,X1,  type="l", ylim=c(0,max(2,X1,X2,X3))),
```

```
   ylab = "Concs")
> lines(t,  X2,  lty=2)
> lines(t,  X3,  lty=3)
> lines(t,  X4,  lty=4)
> legend(6,max(2,X1,X2,X3),legend=c("X1","X2","X3","X4"),
  lty=1:4, bty="n")
```

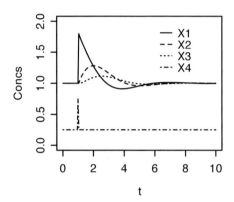

Fig. 9.3 Time-dependence of concentrations for feedback mechanism 9.9)

9.3.2 Transitions in reaction network behavior

As we noted in the previous chapter, networks of chemical and biochemical reactions can lead to oscillating or unstable behavior for certain values of the parameters. An example [63, p. 58] is the system

$$\frac{dX_1}{dt} = \alpha(X_1^4 - X_1^3 X_2^2)$$
$$\frac{dX_2}{dt} = X_1^3 - X_1^2 X_2 \qquad (9.10)$$

with $X_1(0) = X_2(0) = 1$. The system behavior varies greatly depending on whether α is less than or greater than 1. Here is the R code using lsoda to solve the differential equations and the R function approxfun to apply a pulse. Look up approxfun and its close relation approx in R Help, and note their close relation to splinefun and spline.

```
> # A simple S-system with impulse
```

```
> # (Torres and Voit, p. 65)
> library(odesolve)
> lvmodel = function(t, x, parms) {
  X1 = x[1] # Substrate
  X2 = x[2] # Producer
  with(as.list(parms), {
    import = sigimp(t)
    dX1 = a*(X1^4 - X1^3 * X2^2) + import
    dX2 = X1^3 - X1^2 * X2
    res<-c(dX1, dX2)
    list(res)
  })
}

> # Vector of timesteps
> times   = seq(0, 100, length=101)

> # External signal with point impulse
> signal = as.data.frame(list(times = times,
  import = rep(0,length(times))))

> signal$import[signal$times == 50] = 0.5

> sigimp=approxfun(signal$times,signal$import,rule=2)

> # Parameters for steady-state conditions
> parms   = c(a=0.8)

> # Start values for steady state
> y = xstart = c(X1=1, X2=1)

> # lsoda with fixed maximum time step
> out1  = as.data.frame(lsoda(xstart, times, lvmodel,
  parms, atol=1e-4, rtol=1e-4, hmax = 1))

> par(mfrow=c(1,2))
> plot (out1$time, out1$X1,  type="l", ylim=c(0,2),
  xlab="time", ylab="Concs")
> lines(out1$time, out1$X2, lty=2)
> plot (out1$X1,out1$X2,type="l",xlab="X1",ylab="X2")
```

We see that with $\alpha = 0.8$, when subjected to a perturbation the system returns to the original steady state after some damped oscillations. This result holds for any value of $\alpha < 1$. On the other hand, if $\alpha > 1$, the result is oscillations about an unstable steady state (see Problems). An additional example of dramatically changing behavior with change in a parameter is shown in the last problem.

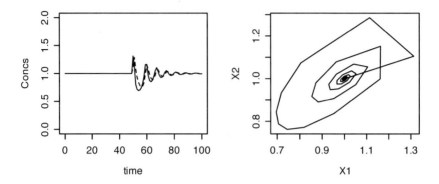

Fig. 9.4 Time dependence and phase plot for mechanism 9.10

9.4 Problems

1. Calculate and plot the fluxes of reactions $S \rightleftharpoons P$ and $P \rightleftharpoons Q$ in section 9.1.2 as functions of time. How do they compare to the input J?
2. Increase one of the V_{max} values 10-fold in the example in Section 9.1.3; what is the resultant increase in flux?
3. Calculate the disequilibrium ratios for the successive 1,1,1 mechanism (Section 9.1.3) with the given pinned values of X,Y.
4. Numerically check the validity of eq. 9.8 for the connectivity theorem.
5. Determine from the eigenvalues of the linearized S-system equation 9.9 whether the steady state is stable.
6. Calculate and plot the concentrations in the example in Section 9.3.1 when X4.off = 10. (A step function rather than a pulse.) Compare the new steady-state values of X1, X2, X3 for the pulse and step perturbations.
7. Run the code in Section 9.3.2 for $\alpha = 1.2$ and a maximum time of 1000, and plot the results including a phase plot. What happens when $\alpha = 1$?
8. The oxidation of NADH catalyzed by peroxidase in the presence of dissolved oxygen
$$2\,NADH + 2\,H^+ + O_2 \rightarrow 2\,NAD^+ + 2\,H_2O$$
was modeled by Steinmetz and Larter [59] (see [63, p. 67]) by the four-step mechanism

$$\frac{d[O_2]}{dt} = 0.89 - k[O_2][NADH][X] - 0.046875[O_2][NADH][Y] - 0.1175[O_2]$$

$$\frac{d[NADH]}{dt} = 0.5 - k[O_2][NADH][X] - 0.046875[O_2][NADH][Y]$$

$$\frac{d[X]}{dt} = 0.01 + k[O_2][NADH][X] + 0.09375[O_2][NADH][Y] - 20[X] - 2500[X]^2$$

$$\frac{d[Y]}{dt} = 2500[X]^2 - 0.046875[O_2][NADH][Y] - 1.104[Y] \qquad (9.11)$$

where X is postulated to be the NAD· radical and Y a radical form of the per-oxidase enzyme. The rate constant k is a parameter which produces markedly different behavior as it is varied. Solve this set of differential equations with k values of 1.0, 0.2, and 0.1 over a range of time from 0 to 500. Show $[O_2]$ and [NADH] as functions of time, and the phase plot of $[O_2]$ vs. [NADH], for each k. Comment on the results.

Chapter 10
Models of Regulation

The functioning of living systems depends on proper regulation of the multiplicity of pathways in the cell and between cells. Whether it be determination of the right time for DNA replication and cell division, transcription of DNA into RNA to produce the proteins needed for various life-cycle stages and for reaction to environmental cues, response to external and internal signals, or development from unicellular to multicellular stages of an organism, proper regulation of a myriad of chemical reactions and noncovalent interactions is essential to life. Much of modern molecular and cellular biology has been devoted to understanding regulation.

In this chapter we shall consider models of regulation in three different types of biological processes: transcription, response to chemotactic signals, and patterning of morphogens in cellular development. These are each huge topics, and we shall attempt only to present some introductory but instructive examples. We shall examine some relatively simple but useful regulatory mechanisms that are amenable to numerical simulation. An important theme will be *robustness*, the ability of a system to maintain suitable functioning in the face of variations, both temporal and cell-to-cell, of biochemical parameters. The book by Alon [3] is an insightful, readable resource for these matters.

10.1 Regulation of transcription: Feed-forward loops

A cell, be it prokaryotic like *E. coli* or eukaryotic like yeast, has thousands of genes whose transcription into proteins is controlled by many hundreds of transcription factors. This control is not generally exerted by one transcription factor acting on one gene, but rather by an intricate, highly connected network. Making sense of this complexity is difficult, but it has been shown (see [44, 3] for reviews) that transcription networks are largely made up of a relatively small number of patterns called "network motifs". In this section we shall consider one of the most common of these motifs, the feed-forward loop.

V. Bloomfield, *Computer Simulation and Data Analysis in Molecular Biology and Biophysics,*   191
Biological and Medical Physics, Biomedical Engineering, DOI: 10.1007/978-1-4419-0083-8_10,
© Springer Science + Business Media, LLC 2009

The feed-forward loop (FFL) consists of two transcription factors, denoted generically X and Y. X regulates Y, and both X and Y regulate the gene Z. There are two classes of FFL, with the most common example of each type diagrammed in Figure 10.1.

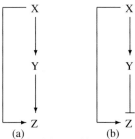

(a)   (b)

Fig. 10.1 Coherent (a) and incoherent (b) feed-forward loops

In a coherent FFL, Figure 10.1 (a), the sign of the direct regulation of Z by X is the same as that of the indirect regulation of Z by X through Y. As diagrammed, both the direct and indirect regulations are activations, indicated by the arrows. In an incoherent FFL, Figure 10.1 (b), the direct and indirect regulations have different signs: the direct regulation is an activation, but the indirect regulation is a repression, indicated by the tee. That is, X activates Y, but Y represses Z. Since there are three regulatory processes in an FFL, and each can be either activation or repression, there are $2^3 = 8$ possible FFLs. The two diagrammed are the most common in both *E. coli* and yeast. We shall see that they have different kinetic consequences and therefore likely different functional roles.

There are two input signals or inducers, S_X and S_Y, that bind to X and Y to produce the active forms X^* and Y^*. We assume that X is constitutively produced at concentration 1. The equations for Y and Z are [44]

$$\frac{dY}{dt} = B_y + \beta_y f(X^*, K_{xy}) - \alpha_y Y \tag{10.1}$$

$$\frac{dZ}{dt} = B_z + \beta_z G(X^*, K_{xz}, Y^*, K_{yz}) - \alpha_z Z \tag{10.2}$$

B_y and B_z are the basal rates of synthesis of Y and Z, β_y and β_z are rate coefficients, and α_y and α_z are the decay rates due to both protein degradation and dilution by cell division.

If production of Z is turned off at $t = 0$, its concentration will decay exponentially: $Z(t) = Z(0) \exp(-\alpha_z t)$. The time to decline to half its initial concentration is defined as the lifetime of Z, equal to $\ln(2)/\alpha_z$. Similar considerations apply to Y.

For an activator u, the f function is

$$f(u,K) = \frac{(u/K)^H}{1+(u/K)^H} \qquad (10.3)$$

and for a repressor

$$f(u,K) = \frac{1}{1+(u/K)^H} \qquad (10.4)$$

K_{ij} is the "binding constant" for activation or repression of gene j by transcription factor i, and H is a Hill coefficient. If both X and Y are required to activate Z, the motif is called an "AND gate", with regulatory function

$$G_z = f(X^*, K_{xz}) f(Y^*, K_{yz}) \qquad (10.5)$$

while if either X or Y can activate Z, we speak of an "OR gate". We shall deal here only with AND gates; for other possibilities see [44].

The behavior of these FFLs can be compared with that of a simple regulatory circuit in which Y is constitutively expressed ($Y = 1$) while differential equation 10.2 for Z remains the same.

In Problems 1 and 2 you are asked to compute the responses of the coherent and incoherent FFLs—both employing AND logic—to abruptly turning on, and then turning off, the signal S_X, and to compare those responses to that of a simple circuit. The result is that the coherent FFL is slower to turn on than the simple circuit, but just as fast to turn off; while the incoherent FFL is faster to turn on. These responses may be useful in buffering the system to fluctuations in nutrients or stress factors which are represented by S_X.

10.2 Regulation of signaling: Bacterial chemotaxis

Bacteria such as *E. coli*, like other organisms, find it useful to be able to move toward nutrients (attractants) and away from toxins (repellants). Remarkably, *E. coli* can sense and swim toward attractant concentration changes as small as one molecule per cell volume per micron, and can do so over a concentration range of five orders of magnitude. What is more, they can do so while subject to Brownian motion bombardment that can reorient them by 90° every ten seconds. The topic is engagingly explored in the book by Berg [7] and in [3, Ch. 7].

E. coli is so small, only a micron or so in length, that it could not detect a concentration gradient by sensing the concentration difference between its ends. A typical attractant concentration might be 1000 molecules per cubic micron, and fluctuation theory shows that this number might vary by $\sqrt{1000} \approx 30$ molecules. One extra molecule at the front compared to the back would be drowned out by that "noise". Instead, the bacterium converts a spatial gradient to a temporal gradient, in which it decreases the frequency of "tumbles" that randomize its direction, thereby increasing the length of time devoted to straight-line runs which on average are directed up the gradient. In the absence of a gradient, runs typically last about 1 sec and are

interrupted by tumbles of duration about 0.1 sec. The runs and tumbles are generated by different states of the flagellar motors that propel the bacterium. These states
are generated in turn by interaction with a bacterial protein kinase that is activated,
by a mechanism we shall shortly examine, through interaction of the attractant or
repellant with receptors on the bacterial cell surface.

10.2.1 Modeling of chemotaxis as a biased random walk

The following code, which you are asked to exercise in Problems 3 and 4, simulates
the biased random walk of a chemotactic bacterium in an attractant solution in which
the attractant concentration increases upstream.

```
> b = 0.01 # Step length
> dtheta = pi/30 # Random angle change on each step
> x0 = 0 # Starting position
> p0 = 0.1 # Prob of tumbling if moving upstream
> pr = 0.99 # Prob of tumbling if moving downstream
> n = 100 # Number of steps
> x = x0 # Initialize position
> # Choose a random starting orientation
> theta = runif(1,0,2*pi)
> dx = b*cos(theta) # x-position change on first step
> for (i in 2:n) {
   x[i] = x[i-1] + dx # New x position
   # Adjust tumbling probability depending on whether
   # it moved up or down
   if (x[i] >= x[i-1]) ptu = p0 else ptu = pr
   r = runif(1)
   # If r > tumbling probability it will not tumble
   if (r >= ptu) {
     # Small rotation to left or right?
     rtheta = runif(1)
     if (rtheta <= 0.5) theta = theta - dtheta
       else theta = theta + dtheta
     # If up continue in direction with small rotation
     dx = b*cos(theta)}
   else { # r < tumbling probability so it will tumble
     # If down then randomize direction
     theta = runif(1,0,2*pi)
     # x-position change after randomizing
     dx = b*cos(theta)
     }
   }
> plot(1:n,x,type="l")
```

10.2.2 Robust model of chemotaxis

At the molecular and cellular level, a mechanism for bacterial chemotaxis that accounts for many of the observed properties, notably the insensitivity (robustness) of tumbling frequency to ligand concentration in a uniform solution, is the following [5].

- A ligand L (attractant or repellant) binds to a receptor on the surface of the bacterium.
- The receptor, denoted MCP, forms stable complexes with two proteins: CheW and CheA. CheW is an adaptor protein that facilitates the binding of CheA to MCP. We denote the entire MCP-CheW-CheA receptor complex as X.
- CheA is a kinase that, as a component of X, phosphorylates a response regulator protein CheY when X is in its active state. The phosphorylase CheZ dephosphorylates CheY.
- The phosphorylated form of CheY binds to the flagellar motor and generates tumbling.
- X is methylated by CheR (abbreviated R in what follows) and demethylated by CheB (abbreviated B). X can have varying degrees of methylation: X_0, X_1, X_2, etc.
- The methylated forms of X, but not the unmethylated X_0, interconvert rapidly between inactive (X) and active (X^*) states. The kinase activity of X^* increases with its extent of methylation. Thus the average kinase activity of the receptor complex, and thereby the tumbling frequency, is enhanced by increasing methylation.
- However, CheA also phosphorylates CheB, thereby increasing its demethylation activity. CheB demethylates only the active form of X. Thus the state of methylation, and kinase activity, of X is controlled by a feedback loop. This feedback is the basis of robust adaptation.

In the model proposed by Barkai and Leibler [5], the receptor X and receptor-ligand complex XL have a maximum of M positions for potential methylation by enzyme R, leading to species denoted X_0, X_1,..., X_M and X_0L, X_1L, ..., X_ML. The affinity of receptor for ligand does not depend on the states of methylation or activation. The methylated species convert rapidly between inactive and active (X_1^*, X_2^*,..., X_1^*L, X_2^*L,...) forms. Binding of ligand by receptor decreases the probability of activation. The ligand association-dissociation reactions and the interconversion between inactive and activated forms are rapid compared to methylation and demethylation, and can be treated as being in equilibrium. The demethylating enzyme B works only on the active conformations. The activity A is defined as the sum of X_1^*, X_2^*,... and their ligand-bound forms. The translation of A into tumbling depends on the kinetics of phosphorylation and dephosphorylation of CheY and on the interaction of CheY with the flagellar motors; these steps are not included in the model [5]. This model is robust with respect to activity for a given set of parameters, showing exact adaptation as ligand is varied. On the other hand, it is not robust with respect to adaptation time.

The reactions and rate equations are

1. ligand binding with dissociation constant K the same for all complexes ($m = 0, 1, \ldots, M$)

$$X_m + L \rightleftharpoons X_mL, \qquad K = \frac{[X_m][L]}{[X_mL]} \tag{10.6}$$

$$X_m^* + L \rightleftharpoons X_m^*L, \qquad K = \frac{[X_m^*][L]}{[X_m^*L]} \tag{10.7}$$

2. reversible association-dissociation of methylating enzyme CheR with receptors ($m = 0, 1, \ldots, M - 1$)

$$R + X_m \underset{d_r}{\overset{a_r}{\rightleftharpoons}} RX_m, \qquad -\frac{d[R]}{dt} = -\frac{[dX_m]}{dt} = \frac{d[RX_m]}{dt} = a_r[R][X_m] - d_r[RX_m]$$
$$\tag{10.8}$$

$$R + X_mL \underset{d_r}{\overset{a_r}{\rightleftharpoons}} RX_mL, \qquad -\frac{d[R]}{dt} = -\frac{[dX_mL]}{dt} = \frac{d[RX_mL]}{dt} = a_r[R][X_mL] - d_r[RX_mL]$$
$$\tag{10.9}$$

3. methylation of X and XL in complex with CheR ($m = 0, 1, \ldots, M - 1$)

$$X_mR \overset{k_r}{\rightarrow} X_{m+1} + R, \qquad -\frac{d[X_mR]}{dt} = \frac{d[X_{m+1}]}{dt} = \frac{d[R]}{dt} = k_r[X_mR] \tag{10.10}$$

$$X_mLR \overset{k_r}{\rightarrow} X_{m+1}L + R, \qquad -\frac{d[X_mLR]}{dt} = \frac{d[X_{m+1}L]}{dt} = \frac{d[R]}{dt} = k_r[X_mLR]$$
$$\tag{10.11}$$

4. rapid transition between inactive and active states with probabilities α that depend on the number of methyl groups and on whether or not the receptor is bound to ligand ($m = 1, 2, \ldots, M$)

$$X_m \rightarrow X_m^*, \qquad [X_m^*] = \alpha_m[X_m]_{tot} \tag{10.12}$$

$$X_mL \rightarrow X_m^*L, \qquad [X_m^*L] = \alpha_m^L[X_mL]_{tot} \tag{10.13}$$

5. reversible association-dissociation of the demethylating enzyme CheB with the activated receptors ($m = 1, 2, \ldots, M$)

$$B + X_m^* \underset{d_b}{\overset{a_b}{\rightleftharpoons}} BX_m^*, \qquad -\frac{d[B]}{dt} = -\frac{[dX_m^*]}{dt} = \frac{d[BX_m^*]}{dt} = a_b[B][X_m^*] - d_b[BX_m^*]$$
$$\tag{10.14}$$

$$B + X_m^*L \underset{d_b}{\overset{a_b}{\rightleftharpoons}} BX_m^*L, \qquad -\frac{d[B]}{dt} = -\frac{[dX_m^*L]}{dt} = \frac{d[BX_m^*L]}{dt} = a_b[B][X_m^*L] - d_b[BX_m^*L]$$
$$\tag{10.15}$$

6. and, finally, demethylation of the active receptors ($m = 1, 2, \ldots, M$)

$$BX_m^* \rightarrow X_{m-1} + B, \qquad -\frac{d[BX_m^*]}{dt} = \frac{d[X_{m-1}]}{dt} = \frac{d[B]}{dt} = k_b[BX_m^*] \quad (10.16)$$

$$BX_m^*L \rightarrow X_{m-1}L + B, \qquad -\frac{d[BX_m^*L]}{dt} = \frac{d[X_{m-1}L]}{dt} = \frac{d[B]}{dt} = k_b[BX_m^*L]$$
$$(10.17)$$

In this simple model, it is assumed that R and B act on their substrates with rates independent of number of methyl groups attached.

In Problems 5–7, you are asked to combine these rate equations to get a set of simultaneous differential equations for the species concentrations, to numerically solve the set of equations, and to draw conclusions about the robustness and other aspects of the mechanism.

10.3 Regulation of development: Morphogenesis

As organisms develop from their initial unicellular state through a sequence of multicellular states, cells that carry identical genetic information must develop distinct roles based on their positions in the growing cell mass. This patterning, or set of cell fate decisions, appears often to be controlled by long-range gradients of morphogen molecules, an idea first put forth by Wolpert [68]. In this model, a morphogen, which is generally a protein, will induce cell fate 1 if it is at or above concentration M_1, fate 2 if it is between M_1 and a lower concentration M_2, fate 3 if it is between M_2 and a still lower concentration M_3, and so on. The induction occurs through binding of M to specific receptors R on the cell surface, with affinities $R_1 < R_2 < R_3 \ldots$. Binding to a given receptor sets off a chain of intracellular events leading to differentiation. Readable summaries of this approach are the review by Eldar et al. [22] and Chapter 8 in the book by Alon [3].

An alternative or complementary hypothesis is that patterning is controlled by the topology of short-range, cell-to-cell interactions [16, 17, 35]. We shall not consider that model here.

A challenge to all models of developmental morphogenesis is that they should be robust to significant changes in morphogen concentrations and environmental conditions such as temperature and nutrition. The most striking evidence for such robustness is that individuals heterozygous for a given morphogen gene, which therefore produce only half as much of the protein as the homozygous individual, generally develop quite normally. Likewise, morphogenesis generally occurs normally over a range of temperatures, which can change reaction rates and equilibria.

10.3.1 *Exponential morphogen gradients are* **not** *robust*

As an example of a mechanism that is not robust to changes in morphogen concentration, consider the production of a morphogen M at a small, localized spot within the cell mass, its diffusion outward, and its degradation along the way. The pertinent reaction-diffusion equation is

$$\frac{\partial M}{\partial t} = D\frac{\partial^2 M}{\partial x^2} - \alpha M \tag{10.18}$$

where M is the morphogen concentration, D the diffusion coefficient, and α the degradation rate constant. At the steady state $\partial M/\partial t = 0$. If M at the source is maintained at M_0, and M at an infinite distance from the source = 0 owing to complete degradation, the solution of the ordinary differential equation

$$\frac{d^2 M}{dx^2} = \alpha M \tag{10.19}$$

with those boundary conditions is the exponential

$$M(x) = M_0 e^{-x/\lambda} \tag{10.20}$$

where $\lambda = \sqrt{D/\alpha}$ is the characteristic decay distance.

Suppose that a cell adopts fate 1 if it is exposed to morphogen at concentrations between M_0 and $\frac{1}{2}M_0$, fate 2 if $\frac{1}{2}M_0 < M \le \frac{1}{4}M_0$, and fate 3 if $M < \frac{1}{4}M_0$. It is easy to see that the values of x setting the boundaries between 1 and 2, and 2 and 3, are at $\lambda \ln 2$ and $\lambda \ln 4$, respectively. The following R code (with $M_0 = 1$ and $\lambda = 1$ for convenience) plots the results, with the regions of x set off by vertical dashed lines and the fate regions labeled. The code then reduces the concentration of morphogen at the origin to 75% of its basal value, and shows the new cell fate boundaries by dotted lines. The boundaries have moved significantly to the left, showing that the patterning is not robust to moderate changes in M_0. In fact, note that if we had reduced the origin concentration to $\frac{1}{2}M_0$, the equivalent of going from homozygous to heterozygous with respect to the gene for the morphogen, there would be no region 1 at all.

```
> x = seq(0,2,0.1)
> M = exp(-x)
> plot(x,M,type="l")
> segments(0,0.5,log(2),0.5,lty=2)
> segments(0,0.25,log(4),0.25,lty=2)
> segments(log(2),0,log(2),0.5,lty=2)
> segments(log(4),0,log(4),0.25,lty=2)
> text(0.5,  0.3,  "1")
> text(1.1,  0.3,  "2")
> text(1.5,  0.3,  "3")
```

```
> M1 = 0.75*exp(-x)  # Reduce M0 from 1 to 0.75
> lines(x,M1,type="l",lty=3)
> segments(log(1.5),0,log(1.5),0.5,lty=3)
> segments(log(3),0,log(3),0.25,lty=3)
```

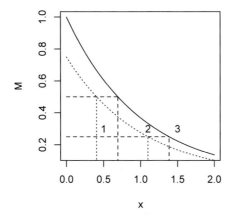

Fig. 10.2 Demonstration that an exponential morphogen gradient is not robust

10.3.2 *Self-enhanced morphogen degradation to form robust gradients*

Mechanisms for achieving a robust steeper-than-exponential gradient of morphogen has been explored by Eldar et al. [21]. They are used for patterning of the wing disc of the fruit-fly *Drosophila*.

We have seen that an exponential distribution of morphogen does not lead to robust pattern formation. The morphogen gradient must decrease more rapidly than exponential near its source, while at the same time extending a considerable distance from the source to accomplish long-range patterning. These seemingly antithetical requirements are reconciled by a mechanism in which morphogens enhance their own degradation, thereby producing a higher degradation rate near the source. In fact, two mechanisms for self-enhanced degradation have been proposed, both involving interactions between the morphogen and its receptor.

In the "passive" mechanism, morphogen is degraded mainly by receptor-mediated endocytosis. At the same time, receptor concentration is up-regulated by morphogen. Thus, the more morphogen the more receptors, and the faster morphogen is removed from regions of high concentration. This passive mechanism is used in differentiation of the wing disc of *Drosophila* with Hedgehog as the morphogen and Patched as the receptor.

In the "active" mechanism, morphogen is primarily degraded by a protease, whose concentration is decreased by complexing with receptor. But morphogen represses the synthesis of repressor. Thus high morphogen concentration leads to less repressor, less sequestering of protease, and more degradation of morphogen. In *Drosophila*, this active mechanism is used to pattern the wing disc with Wingless as the morphogen and Frizzled as the receptor.

A set of reaction-diffusion equations that encompasses both the passive and active mechanisms, with suitable choices of the parameters, is [21]

$$\frac{\partial [Wg]}{\partial t} = D\frac{\partial^2 [Wg]}{\partial x^2} - k_1[DFz2][Wg] + k_{-1}[WgDFz2] - \lambda [Wg][X] \quad (10.21)$$

$$\frac{\partial [DFz2]}{\partial t} = \eta_{Fz}\frac{K^n}{K^n + [WgDFz2]^n} - \alpha[DFz2] - k_1[DFz2][Wg] + k_{-1}[WgDFz2]$$
$$- k_2[DFz2][X] + k_{-2}[XDFz2]$$
$$(10.22)$$

$$\frac{\partial [WgDFz2]}{\partial t} = k_1[DFz2][Wg] - k_{-1}[WgDFz2] \quad (10.23)$$

$$\frac{\partial [XDFz2]}{\partial t} = k_2[DFz2][X] - k_{-2}[XDFz2] \quad (10.24)$$

$$[X]_{tot} = [X] + [XDFz2] \quad (10.25)$$

Equations like this can be solved numerically using the finite difference approach developed in Sections 8.1.3 and 8.2.

λ is the rate of destruction of Wg (ligand) by X (protease), and α is the rate of DFz2 (receptor) degradation. $k_{\pm 1}$ are the association and dissociation rates of the WgDFz2 complex, and $k_{\pm 2}$ the equivalent rates for the complex of receptor with protease. K is the signaling affinity, and n a Hill coefficient. η_{Fz} is the receptor production rate, and η_{Wg} is the ligand production rate at the boundary.

In Problems 8 and 9 you are asked to numerically solve these equations and plot the results, given suitable values of the parameters.

10.3.3 Patterning in the dorsal region of Drosophila

A different mechanism for establishing a robust morphogen gradient was elucidated by Eldar and collaborators [20] for patterning in the dorsal region of *Drosophila*. In this case the morphogen is produced uniformly, but is shuttled to the dorsal midline where it builds up a high concentration.

The early stage embryo of the fruit fly *Drosophila* is like a cylinder. In cross-section perpendicular to its long axis, after about 2 hours, it consists of three domains of gene expression: the dorsal region (DR) above and the mesoderm (M) be-

low, separated by two regions of neuroectoderm (NE). These domains are schemat-
ically illustrated in the figure, which results from cutting the circular cross-section
in the middle of the mesoderm and laying it out flat.

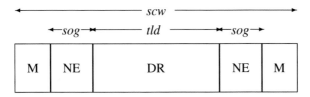

The dorsal region next becomes subdivided or patterned to form the amnioserosa
and the dorsal ectoderm. This subdivision takes place under the influence of a patterning network consisting of the following proteins:

- Ligands of the bone morphogenic protein (BMP) class: Scw and Dpp
- Sog, a BMP inhibitor
- Tld, a protease that cleaves Sog
- Tsg, an accessory protein that must bind to Sog in order that Sog can bind to Dpp

Free Sog is not cleaved by Tld, but Sog in the Scw-Sog complex is cleaved. The
diffusion of free Scw (or Dpp) is restricted, while the Scw-Sog and Dpp-Sog-Tsg
complexes diffuse freely.

We consider a simplified model with only Scw, Sog, and Tld. Diffusional transport of Scw via the Scw-Sog complex to the dorsal midline is the key event in
patterning. After Scw is shuttled by Sog to the dorsal midline, it is then degraded
by Tld. Analysis of this mechanism shows that at distances far from the dorsal midline, the Scw concentration dies off like $1/x^2$, independent of the concentrations
of the other BMP network components. This is the behavior that underlies robustness. "The robustness mechanism relies on the ability to store an excess of signaling
ligand molecules in a restricted spatial domain where Sog is largely absent" [22].

A set of reaction-diffusion equations that describes this mechanism is [20]

$$\frac{\partial [Sog]}{\partial t} = D_S \frac{\partial^2 [Sog]}{\partial x^2} - k_b[Sog][Scw] + k_{-b}[SogScw] - \alpha[Tld][Sog] \quad (10.26)$$

$$\frac{\partial [Scw]}{\partial t} = D_{BMP} \frac{\partial^2 [Scw]}{\partial x^2} - k_b[Sog][Scw] + k_{-b}[SogScw] + \lambda[Tld][SogScw]$$
$$(10.27)$$

$$\frac{\partial [SogScw]}{\partial t} = D_C \frac{\partial^2 [SogScw]}{\partial x^2} + k_b[Sog][Scw] - k_{-b}[SogScw]$$
$$- \lambda[Tld][SogScw] \quad (10.28)$$

In Problem 10 you are asked to solve this set of equations for a set of normalized, dimensionless parameters, then to vary one of the parameters substantially and
compare the results.

10.4 Problems

1. Numerically solve eqs. 10.1–10.3 and 10.5 for the behavior of a coherent FFL obeying AND logic, in the presence of a pulse of inducer S_X. Turn S_X from 0 to 1 at $t = 0$, and turn it back to 0 at $t = 3$, where t is measured in units of the lifetime of Z. Assume that when $S_X = 1$, all X is in the form of X^*, and when $S_X = 0$, $X^* = 0$. Also assume that $S_y = 1$, so that all Y is converted to Y^*. Use the parameters [44] $H = 2$, $\beta_y = \beta_z = 1$, $\alpha_y = \alpha_z = 1$, $B_y = B_z = 0$, $K_{xz} = K_{xy} = 0.1$, and two values of K_{yz}, 0.5 and 5. Plot the results (Z as a function of t), over the time interval $0 < t < 6$. On the same graph, plot the results for simple regulation (Y is constitutively expressed) with parameters $Y = 1$, $H = 2$, $\beta_z = 1$, $\alpha_z = 1$, $B_z = 0$, and $K_{xz} = 1$. You should find that the rise in Z is slower for the FFL than for the simple regulatory circuit, but the decay is the same.

2. Repeat Problem 1 for the incoherent FFL with basal Y activity and AND logic, using eq. 10.4 instead of eq. 10.3 and the following parameters: $K_{xz} = K_{xy} = 1$, $K_{yz} = 0.5$, and two values of B_y, 0.5 and 0.3. All other parameters are the same as in Problem 1. Again compare with the simple regulatory circuit. You should find that Z rises more quickly for the incoherent FFL than for the simple circuit, and overshoots its steady-state level before decaying when S_X is turned off.

3. Modify the code for biased random walk of a chemotactic bacterium in the presence of an attractant gradient in the x-direction to calculate y as well as x values. Then plot y vs x for 100 steps.

4. Write a program to repeat the code for biased random walk of a chemotactic bacterium in the presence of an attractant gradient 500 times, and plot a histogram of the end-points.

5. Combine eqs. 10.6–10.17 to form a set of simultaneous ordinary differential equations for the time-dependence of the various species.

6. Numerically solve the simultaneous ODEs in the previous problem, using values for the parameters from Barkai and Leibler [5]: $K = 1$ μM, $a_r = 80$ $s^{-1}\mu M^{-1}$, $d_r = 100$ s^{-1}, $k_r = 0.1$ s^{-1}, $a_b = 800$ $s^{-1}\mu M^{-1}$, $d_b = 1000$ s^{-1}, $k_b = 0.1$ s^{-1}. Calculate micromolar concentrations from the cell volume (1.4×10^{-15} L) and the number of molecules per cell: 10,000 X, 2000 CheB, and 300 CheR. Set the maximum number of methylation sites $M = 4$. The probabilities that an unliganded receptor with 1...4 methyl groups is in the activated state are $\alpha = (0.1, 0.5, 0.75, 1.0)$, while the probabilities for a liganded receptor are $\alpha^L = (0, 0.1, 0.5, 1)$. Start by setting the ligand concentration $[L] = 1$ μM and letting X, R, and B all be in their uncomplexed forms, and run the calculation until a steady state is reached. Plot the activity A as a function of time.

7. Once the steady state in the previous problem has been reached, abruptly increase the ligand concentration to 10 μM and plot the change in A until a new steady state is reached. Is the solution robust with respect to activity? How long does it take for the activity to readjust to 1/2 of its original value (the adaptation time)? Then abruptly decrease $[L]$ to 2 μM and repeat the calculation. What are the new values of the activity and adaptation time?

8. Numerically solve eqs. 10.21–10.25 with the following passive stabilization parameters in the region $0 < x < L_p$: $L_p = 4.5$ μm, $D = 0.1$ μm^2/s, $k_1 = 6.7 \times 10^{-3}$ μM^{-1}s^{-1}, $k_{-1} = 3.3 \times 10^{-3}$ s^{-1}, $\lambda = 3.7 \times 10^{-2}$ μM^{-1}s^{-1}, $\eta_{Wg} = 2.4 \times 10^{-3}$ μMs^{-1}, $\eta_{Fz} = 1.7 \times 10^{-3}$ μMs^{-1}, $K = 2 \times 10^{-3}$ μM, $n = 1.5$, $\alpha = 3.3 \times 10^{-4}$ s^{-1}, $k_2 = 0$ μM^{-1}s^{-1}, $k_{-2} = 3.3 \times 10^{-4}$ s^{-1}, and $[X]_{tot} = 0.3$ μM. The flux at $x = 0$ is $\eta_{Wg}L_p$, while the flux at $X = L_p$ is 0. Use the finite-difference approach developed in Sections 8.1.3 and 8.2 to combine diffusion and reaction steps.

9. Then for active stabilization redo the calculation with $k_2 = 0.16$ μM^{-1}s^{-1}, $\eta_{Wg} = 4.4 \times 10^{-2}$ μMs^{-1}, and $\lambda = 1.4 \times 10^{-2}$ μM^{-1}s^{-1}. Use η_{Fz}/α as the initial concentration of DFz2.

10. Solve eqs. 10.26–10.28 in the region $-1 < x < 1$, with the parameters $D_S = 1$, total Scw concentration $[Scw]_{av} = 1$, $D_{BMP} = 0.1$, $D_C = 1$, $k_b = 10$, $k_{-b} = 1$, $\lambda[Tld] = 10$, $\alpha[Tld] = 10$, and a constant flux of Sog into the dorsal region from the NE-DR boundaries of $\eta_S = 10$. Plot the resulting concentration of Scw vs. x. Then change one of the parameters by a factor of 10, redo the calculation, and plot the result on the same graph.

Part III
Analyzing DNA and Protein Sequences

Chapter 11
Probability and Population Genetics

Up to now we have mainly treated deterministic processes, although we have showed how to fit noisy data and to model stochastic chemical reactions. In fact, most biological data are intrinsically noisy, or random, due to the underlying nature of the process (e.g., mutation or genetic recombination), especially when combined with the often small numbers of "individuals" in many experiments. In this chapter, we discuss basic concepts of randomness and probability, and show how they may be applied in a variety of situations, concluding with a brief introduction to population genetics. A good basic treatment of probability in a biological context is the book by Adler [1].

11.1 Some fundamentals of probability

11.1.1 Review of basic probability ideas

We review some of the basic concepts of probability, with which you should already be familiar.

Sample space: the set of all possible results. For flipping a coin, the sample space is H and T. For rolling a die, it is the integers from 1 to 6. For a uniformly distributed random number generated by R using `runif`, it is all the floating point numbers between 0 and 1.

Simple event: Any particular outcome. H or T; 1, 2, 3 ,4, 5, or 6; etc. An **event** is a set of one or more simple events, or a subset of the sample space.

Four requirements of a probability model: Let $Pr(A)$ stand for the probability of A.

- If S is the sample space, $Pr(S) = 1$.
- $0 \le Pr(A) \le 1$ for any event A.
- If A and B are mutually exclusive events, $Pr(A \text{ or } B) = Pr(A) + Pr(B)$.

V. Bloomfield, *Computer Simulation and Data Analysis in Molecular Biology and Biophysics,*   207
Biological and Medical Physics, Biomedical Engineering, DOI: 10.1007/978-1-4419-0083-8_11,
© Springer Science + Business Media, LLC 2009

- If A' is the complement of A, then $Pr(A') = 1 - Pr(A)$.

If the sample space contains a finite number n of simple events A_i,

$$\sum_{i=1}^{n} Pr(A_i) = 1 \tag{11.1}$$

If each simple event i has associated with it a value x_i, the expectation (mean value) of x is

$$E(x) = \sum_{i=1}^{n} x_i Pr(A_i) \tag{11.2}$$

The extension of these ideas to continuous distributions through distribution functions will be discussed later in this chapter.

11.1.2 Conditional probability

Conditional probability is the probability of one event conditional on knowing that another occurred. The mathematical expression is

$$Pr(A|B) = \frac{Pr(A \cap B)}{Pr(B)} \tag{11.3}$$

which is to be read: The probability of A conditional on B is the probability of both A and B divided by the probability of B. The $\cap$ symbol stands for "intersection" (both A and B). You are probably also familiar with "union", denoted by $\cup$, which means either A or B.

Consider a hypothetical example from mouse genetics. Suppose mouse coat color is governed by a single dominant allele W (W = white, w = black). Two mice of genotype Ww (created by crossing white and black inbred mice) are crossed. We are given a normal offspring from this cross. What is the probability that its genotype is WW? The probability of WW is 1/4, and the probability of white is 3/4. (WW, Ww, and wW will be white, ww will be black.) Therefore,

$$Pr(WW|Normal) = \frac{Pr(WW \cap Normal)}{Pr(Normal)} = \frac{1/4}{3/4} = \frac{1}{3} \tag{11.4}$$

11.1.3 The law of total probability

Let X_1, X_2, ..., X_n be a set of mutually exclusive and collectively exhaustive events. Then for every event A

$$Pr(A) = \sum_{i=1}^{n} Pr(A|X_i)Pr(X_i) \tag{11.5}$$

As an example of the law of total probability, suppose that the storeroom stocks an enzyme that you use in your research. They order from three manufacturers: 50% from M_1, 30% from M_2, and 20% from M_3. Occasionally the enzyme is inactive: 3% of the time from M_1, 4% from M_2, and 5% from M_3. If your order is filled at random from one of the three manufacturers, what is the probability $Pr(D)$ that you will receive a defective enzyme?

We can use the law of total probability to find $Pr(D)$ as

$$Pr(D) = Pr(M_1)Pr(D|M_1) + Pr(M_2)Pr(D|M_2) + Pr(M_3)Pr(D|M_3) \tag{11.6}$$
$$= 0.03 \times 0.5 + 0.04 \times 0.3 + 0.05 \times 0.2 = 0.037$$

There is a 3.7% chance that your enzyme will be inactive.

The R code to do this calculation, taking advantage of vectorization, is simply

```
> PrDM = c(0.03,0.04,0.05)
> PrManu = c(0.5,0.3,0.2)
> PrD = PrManu*PrDM
> sum(PrD)
[1] 0.037
```

11.1.4 Bayes' theorem

Bayes' theorem allows calculation of the probability of B conditional on A—$Pr(B|A)$—if we know $Pr(A)$, $Pr(B)$, and $Pr(A|B)$. The theorem, which is derived from the equation for conditional probability above, states

$$Pr(B|A) = \frac{Pr(A|B)Pr(B)}{Pr(A)} \tag{11.7}$$

Adler [1, pp. 519–520] gives a couple of instructive applications. In the first, suppose we know that a person is male with probability $Pr(M) = 0.5$, that males are colorblind with probability $Pr(C|M) = 0.05$, and that all people are colorblind with probability $Pr(C) = 0.03$. Then the probability $Pr(M|C)$ that a colorblind person is male is

$$Pr(M|C) = \frac{Pr(C|M)Pr(M)}{Pr(C)} = \frac{0.05 \times 0.5}{0.03} \approx 0.833 \tag{11.8}$$

An application that often arises in practice is testing for a rare disease. Suppose that the disease afflicts only 1% of the population, i.e., $Pr(D) = 0.01$, so that the probability that a person does not have the disease is $Pr(N) = 0.99$. A test always detects (shows positive, P) those who have the disease: $Pr(P|D) = 1.0$; but gives 5% false positives: $Pr(P|N) = 0.05$. What is the probability that a person who tests positive actually has the disease?

$$Pr(D|P) = \frac{Pr(P|D)Pr(D)}{Pr(P)} \tag{11.9}$$

We don't know $Pr(P)$, but it can be calculated from the law of total probabilities:

$$Pr(P) = Pr(P|D)Pr(D) + Pr(P|N)Pr(N) \tag{11.10}$$
$$= 1.0 \times 0.01 + 0.05 \times 0.99 = 0.0595$$

Then the Bayes calculation can be concluded:

$$Pr(D|P) = \frac{Pr(P|D)Pr(D)}{Pr(P)} = \frac{1.00 \times 0.01}{0.0595} \approx 0.168 \tag{11.11}$$

Although this is quite an accurate test, an individual drawn from the full population who tests positive is still unlikely to have the disease. Therefore, testing is often limited to high-risk groups. You can verify that if, within such a group, there is a 40% chance of having the disease, the probability that a positive test indicates the presence of the disease is 0.93.

The combination of eqs. 11.7 (Bayes' theorem) and 11.5 (the law of total proba-bility) yields a general statement of Bayes' formula:

$$Pr(X_k|A) = \frac{Pr(A|X_k)Pr(X_k)}{\sum_{i=1}^{n} Pr(A|X_i)Pr(X_i)} \tag{11.12}$$

If we think of the events X_k as possible causes of the event A, then Bayes' formula allows us to calculate the probability that a particular one of the X_ks occurred given that A occurred.

11.2 Stochastic population models

We use the term "stochastic" to describe unpredictable events. A familiar stochastic process is the tossing of a coin. On each toss, either H or T comes up, with equal probabilities = 1/2 if it is a fair coin. The result of a given toss does not depend on the result of the previous toss, so this is a stochastic process with no "memory". The use of random variables to generate a result is sometimes called the "Monte Carlo method", after the famed European gambling casino.

11.2.1 *Stochastic modeling of population growth*

As an example of a biologically relevant stochastic process with no memory, we consider population growth under conditions that the growth rate varies unpredictably around some mean value from generation to generation. An R function that enables this calculation is

```
> bn = function(n,b0, mu, sd) {
  b = rep(0,n+1); b[1] = b0;
   for (i in 2:(n+1)) {
  b[i] = trunc(b[i-1]*rnorm(1, mu,sd))}
  return(b)}
```

where n is the number of generations, b0 is the starting population, b[i] is the population in generation i, mu is the mean per capita reproduction rate (r in previous treatments), sd is the standard deviation of the normally distributed values of the reproduction rate, the trunc() function rounds to the closest integer toward 0, and bn is the resulting population vector.

Each calculation of bn will give a different result if sd is greater than zero. For example, consider starting with a population of 100 and running for 50 generations with a mean growth rate of 1.1 (10% increase per generation). If sd = 0, we get the expected exponential growth (solid line). However, two simulations with sd = 0.2 (a roughly 20% variability on either side of the mean) give quite different results, with the population falling sometimes above, sometimes below, the deterministic curve.

```
> x = 0:50 # Generations
> y1 = bn(50,100,1.1,0)
> y2 = bn(50,100,1.1,.2)
> y3 = bn(50,100,1.1,.2)
> plot(x,y1, type="l", lty=1, xlab="Generations",
  ylab="Population")
> lines(x,y2, type="l",lty=2);
> lines(x,y3, type="l",lty=3)
```

Plotting the data in semi-log fashion shows strong deviations from nonlinearity for y2 and y3.

```
> plot(x,log(y1),type="l", lty=1, xlab="Generations",
  ylab="Log(Population)")
> points(x,log(y2),pch=2); points(x,log(y3),pch=3)
```

Fitting the log data to a straight line using lm(log(y) x) gives slopes of 0.947, 0.0957, and 0.0762 for y1, y2, and y3. These are all in the expected range of 10% increase per generation, but y2 and y3 are significantly different from the "noiseless" value and from each other.

If the starting population is small, there is a significant chance that the population will be extinguished due to random fluctuations. Here is an example with n0 =

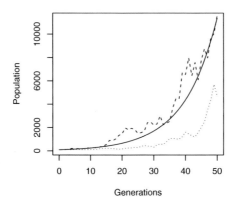

Fig. 11.1 Stochastic and deterministic modelsl of population growth

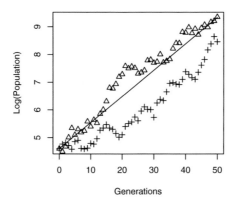

Fig. 11.2 Semi-log plot of stochastic and deterministic models of population growth

10. This is also a possibility, even with larger n0, if sd is large, so there are large fluctuations in the growth rate due to environmental changes.

```
> y2 = bn(50,10,1.1,0.2)
> y3 = bn(50,10,1.1,0.2)
> plot(x,y2, type="l", lty=1, xlab="Generations",
  ylab="Population", ylim=c(0,max(y2)))
> lines(x,y3,lty=2)
```

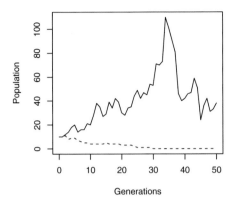

Fig. 11.3 Examples of large fluctuations in stochastic population growth, in one case to extinction

11.2.2 Stochastic simulation of radioactive decay

As another example of how to simulate a stochastic process if the system is too small or too random to allow deterministic simulation, consider radioactive decay from a relatively small number of atoms. In a unit time interval, let the probability that an atom decays be p, so the probability that it does not decay $= 1 - p$. The function below starts with b_0 atoms, and moves through n time periods. In each period, it confronts each remaining atom with a random number chosen from a uniform distribution between 0 and 1. If the random number is less than or equal to p, the atom is removed from the system (i.e., b is decreased by 1); otherwise, b remains the same. This test is applied to each atom of that "generation". The process continues through n generations, and then the population vector b (the number of atoms remaining as a function of time) is returned.

```
> rd=function(n,b0,p) {
b = rep(0,n+1); b[1]=b0;
for (i in 2:(n+1)) {
bi = b[i-1]
s = 0 # s counts number of decayed atoms to be removed
for (j in 1:bi) {pp = runif(1);if (pp <= p) {s = s+1}}
b[i] = b[i-1]-s
}
return(b)
}
```

The deterministic version of this simulation says that a fraction $1 - p$ remains after the first generation, $(1 - p)^2$ after the second generation, ... $(1 - p)^x$ after the xth

generation, ... We plot the stochastic points and the deterministic line for comparison, for a system starting with 100 atoms and a decay probability per generation of
0.1, over 50 generations:

```
> y = rd(50,100,0.1)
> x = 0:50
> plot(x,y, xlab="Time", ylab="Atoms")
> lines(x,100*0.9^x)
```

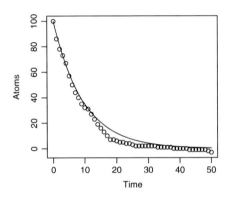

Fig. 11.4 Deterministic and stochastic models of radioactive decay

11.3 Markov chains

Sometimes the probability of finding a given state depends on the previous state.
This is called a "Markov chain", which behaves as if it has some memory of the
previous state. Markov chains are important in a variety of biological contexts, such
as DNA sequence analysis and cooperative binding or conformational transitions
in biopolymers. In this section we show a simple application of the Markov chain
concept. In the next chapter we develop Markov chain ideas in more detail.

11.3.1 Markov chain model of diffusion out of and into a cell

We consider a model from Adler [1, pp. 497–498]. Suppose that there are initially
b_{in} molecules inside a cell, and b_{out} molecules outside. In a unit time interval, the

probability that an "inside" molecule will diffuse out is p_i, and the probability that an "outside" molecule will diffuse in is p_o. Thus the state of the system at time i depends on the state at the previous time, which makes this a Markov chain.

An analytical solution to this problem can be obtained as follows. Let p_t be the probability that a given molecule is inside at time t, so the probability that it is outside is $1 - p_t$. There are two ways in which a molecule can be inside at time $t + 1$: It can be inside at time t, and stay inside with probability $(1 - p_i)$, or it can be outside at time t and diffuse inside with probability p_o. Thus

$$p_{t+1} = (1 - p_i)p_t + p_o(1 - p_t) = (1 - p_i - p_o)p_t + p_o \tag{11.13}$$

Setting

$$p_{t+1} = p_t + \frac{\Delta p_t}{\Delta t} \approx p(t) + \frac{dp}{dt} \tag{11.14}$$

gives

$$\frac{dp(t)}{dt} = -(p_i + p_o)p(t) + p_o \tag{11.15}$$

which can be integrated to give

$$p(t) = p(0)e^{-(p_i + p_o)t} + \frac{p_o}{p_i + p_o}\left(1 - e^{-(p_i + p_o)t}\right) \tag{11.16}$$

At long time, the system reaches equilibrium at a population of molecules inside the cell equal to $p_o/(p_i + p_o)$.

A stochastic simulation of this model proceeds similarly to the radioactive decay model, except that the total number of molecules remains fixed, while the numbers inside and out change with time.

```
> cd = function(n,b.tot,b0.i,p.i, p.o) {
b.i = rep(0,n+1); b.i[1] = b0.i;
b.o = rep(0,n+1); b.o[1] = b.tot-b0.i
for (i in 2:(n+1)) {
bi = b.i[i-1]; bo = b.o[i-1]; si = 0; so = 0
for (j in 1:bi) {pp = runif(1)
   if (pp <= p.i) {si = si-1; so = so+1}}
for (j in 1:bo) {pp = runif(1)
   if (pp <= p.o) {si = si+1; so = so-1}}
b.i[i] = b.i[i-1]+si; b.o[i] = b.o[i-1]+so
}
return(b.i)
}
```

An R function that allows comparison of the analytical expression above with the stochastic calculation is

```
> ycd.calc = function(x,b.tot,b0.i,p.i, p.o) {
a = p.i+p.o; feq = p.o/(p.i+p.o); p0 = b0.i/b.tot;
return (b.tot*(p0*exp(-a*x) + feq*(1-exp(-a*x)))) }
```

In Problem 4 you are asked to plot the stochastic and analytical solutions on the same graph, for specific values of the parameters.

We shall deal with a more systematic representation of Markov processes, using transition matrices, in the next chapter.

11.4 Probability distribution functions

Many simulations in biology and other branches of science involve drawing samples from a given probability distribution. For example, to choose one member of a large group with equal probability, we use the uniform distribution. To simulate random error of a measurement, we generally use the normal distribution. To calculate the probability that exactly k monomers will link together to form a polymer, given that the probability of forming a monomer-monomer bond is p, we use the geometric distribution. To calculate the probability that a given number of events will occur in a fixed period of time if we know the average rate and the events are independent of each other, we use the Poisson distribution. R has these and many other probability distributions. Calculations with them are performed with a generally uniform syntax:

11.4.1 R's four basic functions for a statistical distribution

d   gives the density: the probability of getting a particular value x (discrete distribution) or a value "close to" x (continuous distribution)

p   gives the cumulative probability distribution function (the area under the curve to the left of x) of getting a value q or less

q   gives the quantile function: the value of x such that there is a probability p of getting a value $\leq x$

r   generates random numbers chosen from the distribution

All of the distributions we will consider have additional parameters that may be needed on occasion, though we will not use them. Typically, they can be seen in the R help system as

```
log, log.p: logical; if TRUE, probabilities p
   are given as log(p).
lower.tail: logical; if TRUE (default), probabilities
   are P[X <= x],
otherwise, P[X > x].
```

11.4.2 Uniform distribution

See ?Uniform in R Help for details of the Uniform Distribution. The default minina and maxima are 0 and 1. The uniform distribution has density $f(x) = 1/(max - min)$ for $min \leq x \leq max$. The calling syntax is

```
dunif(x, min, max) # If no min,max specified, uses 0,1
punif(q, min, max)
qunif(p, min, max)
runif(n, min, max) # n is # of random numbers desired
```

Examples

```
> # Probability density at three values of x
> dunif(c(0,.2,.4),0,1)
[1] 1 1 1
> dunif(c(0,.2,.4)) # Default min and max
[1] 1 1 1
> dunif(c(0,.2,.4),0,3)
[1] 0.3333 0.3333 0.3333
> # The distribution is 3 long, so only 1/3 high
> x = seq(0,2,.1)
> plot(x,dunif(x,0,2),type="l",xlim=c(0,3))
```

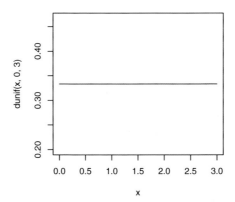

Fig. 11.5 Density function dunif for uniform distribution

```
> # Cumulative probability density at three values of x
> punif(c(0,.2,.4),0,1)
```

```
[1] 0.0 0.2 0.4
> punif(c(0,.2,.4),0,3)
[1] 0.00000 0.06667 0.13333
> x = seq(0,2,.2)
> plot(x,punif(x,0,2),xlim=c(0,2), ylim=c(0,1))
```

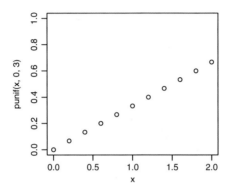

Fig. 11.6 Cumulative density function punif for uniform distribution

```
> qunif(.5,0,1)
[1] 0.5
> x = seq(0,1,.1)
> plot(x,qunif(x,0,5))

> y = runif(10,0,6)
> y
 [1] 3.28621 3.49709 3.49794 0.00719 2.64671 1.87892
     4.44009 0.82996 5.23067 3.13842
> ceiling(y)  # A way to throw 10 dice.
> # See ?round for R rounding functions
 [1] 4 4 4 1 3 2 5 1 6 4

> runif(5) # Default limits (0,1)
 [1] 0.44516 0.05306 0.33621 0.06721 0.63561
> runif(5) # Different random numbers this time
 [1] 0.29592 0.02847 0.18941 0.69989 0.27787

> # To reproduce run of random numbers, set same seed
```

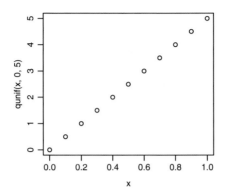

Fig. 11.7 Quantile function `qunif` for uniform distribution

```
> set.seed(234)
> runif(5)
 [1]  0.74562 0.78171 0.02004 0.77609 0.06691
> runif(5)
> # No seed, so different
 [1]  0.555725 0.547702 0.582848 0.582990 0.001198
> set.seed(234)  # Same seed, so the same
> runif(5)
 [1]  0.74562 0.78171 0.02004 0.77609 0.06691
```

11.4.3 Normal distribution

See `?Normal` in R Help for details of the Normal Distribution. The default mean and sd are 0 and 1. The normal distribution has density

$$f(x) = \frac{1}{\sqrt{2\pi}\sigma} e^{-(x-\mu)^2/2\sigma^2}$$
(11.17)

where μ = mean and σ = sd. The calling syntax is

```
dnorm(x, mean, sd)
pnorm(q, mean, sd)
qnorm(p, mean, sd)
rnorm(n, mean, sd)
```

Examples

```
> x = seq(-4,4,0.1)
> plot(x,dnorm(x),type="l")
```

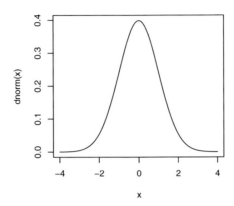

Fig. 11.8 Density function dnorm for normal distribution

```
> dnorm(c(-1,0,1))
[1] 0.2420 0.3989 0.2420

> # Cumulative probability density
> plot(x,pnorm(x),type="l")

> x = seq(0,1,.1)
> plot(x,qnorm(x),type="l")

> # Value of x for which cumulative probability = 0.2
> qnorm(.2)
[1] -0.8416

> # 5 normal random numbers with mean = 10 & sd = 2
> rnorm(5,10,2)
[1] 10.399 10.597   9.362   8.257 12.375

> # 100 normal random numbers with mean = 10 & sd = 2
> y = rnorm(100,10,2)
> # Draw histogram
> hist(y,prob=T,ylab="Probability",
  main="Histogram and Density")
> lines(density(y)) # and add density plot
```

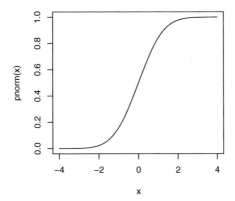

Fig. 11.9 Cumulative distribution function pnorm for normal distribution

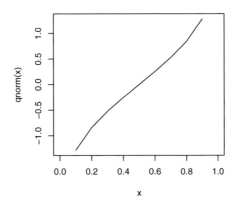

Fig. 11.10 Quantile function qnorm for normal distribution

A biologically pertinent example, after [55, p. 95], is the distribution of restriction fragment lengths when an enzyme acts on a DNA molecule. Suppose that the lengths in base pairs, as measured by gel electrophoresis, are
fraglen=c(278, 281, 290, 295, 297, 302, 306, 309, 310, 313, 317, 320, 324,327, 333, 348, 349, 351, 360, 375)
This appears to be a normal distribution, as tested by qqnorm(fraglen), with mean = 319 and sd = 26.7. Then, to obtain the probability of finding a fragment bigger than 400 base pairs, use

```
> 1-pnorm(400,mean(fraglen), sd(fraglen))
```

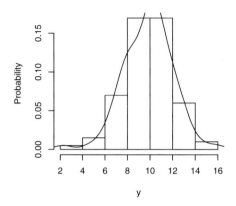

Fig. 11.11 Histogram and density function for 100 normal random numbers with mean = 10 and sd = 2

```
[1] 0.00126
```
to calculate the area to the right of 400 under the cumulative distribution curve.

11.4.4 Binomial distribution

The binomial distribution is the probability distribution of the number of successes in a sequence of n independent yes/no experiments, each of which yields success with probability p. A single success/failure experiment is also called a Bernoulli trial, and when $n = 1$, the binomial distribution is a Bernoulli distribution.

See ?Binomial in R Help for details of the Binomial Distribution. The binomial distribution with size = n and probability = p has density

$$p(x) = \frac{n!}{x!(n-x)!}p^x(1-p)^{n-x}, \qquad x = 0\ldots n \qquad (11.18)$$

The calling syntax is

```
dbinom(x, size, prob) : size = n = number of trials,
   prob = p of success on each trial
pbinom(q, size, prob)
qbinom(p, size, prob)
rbinom(n, size, prob)
```

Examples

```
> par(mfcol=c(1,2))
> x = 0:50
> plot(x,dbinom(x,50,1/3),type="h");
> plot(x,pbinom(x,50,1/3),type="h")
> # h stands for histogram,
> # suitable for discrete distribution
> par(mfcol=c(1,1))
```

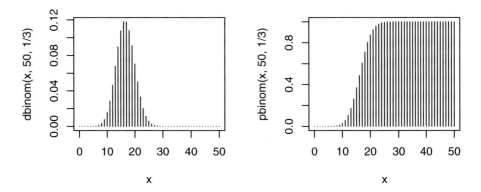

Fig. 11.12 (left) dbinom and (right) pbinom histograms for binomial distribution

```
> x1 = seq(0,1,.01)
> plot(x1,qbinom(x1,50,1/3),type="l")

> # 10 flips of a coin (size = 1) with p(heads)=0.5
> rbinom(10,1,.5)
 [1] 1 0 0 1 1 1 0 1 0 0
```

There are two functions related to the binomial distribution that should be mentioned here. The $n!/[k!(n-k)!]$ term in the binomial probability density distribution stands for the number of combinations of k objects chosen from n; it is sometimes called the "binomial coefficient". It can be calculated in R with the choose function:

```
> choose(40,5) # Ways of choosing 5 objects out of 40
[1] 658008
```

The second function is sample, which chooses k values at random from a list of n. For example, to choose 5 numbers at random from the first 40,

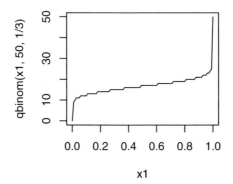

Fig. 11.13 qbinom for binomial distribution

```
> sample(1:40,5)
[1]   3   2 34 22 15
```

The default is to sample without replacement, so that once a number is chosen it can't be chosen again. The other option is to sample with replacement. For example, to simulate the flipping of a coin 10 times:

```
> sample(c("H","T"),10, replace = T)
 [1] "T" "T" "T" "T" "H" "H" "T" "H" "H" "T"
```

11.4.5 Poisson distribution

The Poisson distribution expresses the probability of a number of events occurring in a fixed period of time if these events occur with a known average rate and independently of the time since the last event. The Poisson distribution can also be used to calculate the probability of the number of events in other specified intervals such as distance, area, or volume.

The Poisson distribution is the limiting case of the binomial distribution when the probability of a successful event is low. See ?Poisson in R Help for details. The Poisson distribution for the probability of k events, if the mean value of k is λ, is

$$p(k) = \frac{\lambda^k e^{-\lambda}}{k!} \tag{11.19}$$

The calling syntax is

```
dpois(x, lambda)
```

```
ppois(q, lambda)
qpois(p, lambda)
rpois(n, lambda)
```

Examples

```
> # A Petri dish with an average of 3 colonies
> dpois(x=0:6,lambda=3) # Prob. of 0 to 6 colonies
[1] 0.0498 0.1494 0.2240 0.2240 0.1680 0.1008 0.0504

> ppois(0:6,3) # Cumulative probabilities
[1] 0.0498 0.1991 0.4232 0.6472 0.8153 0.9161 0.9665

> # Prob of .5 that a dish will contain < 3 colonies
> qpois(.5,3)
[1] 3

> hist(rpois(100,3))
> lines(k, (100*3^k*exp(-3))/factorial(k))
```

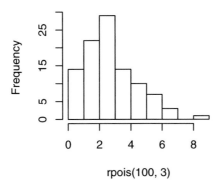

Histogram of rpois(100, 3)

Fig. 11.14 Histogram of the occurrences of 0–6 colonies on 100 plates given a mean of 3 colonies/plate, according to the Poisson distribution

For an additional instructive example, involving the probability of making a given number of errors in a sequencing technique, see [55, p. 87].

11.4.6 Geometric distribution

The geometric distribution, and its continuous version the exponential distribution, give the probability of the first occurrence of an event whose probability is p. It can thus be used to simulate the time of diffusion of a molecule out of a cell, the probability that an end-to-end polymerization will terminate, etc. The distribution function for the probability that the system will proceed for x steps without the event occurring, and then terminate on the $(x+1)$ step, is $(1-p)^x p$. See ?Geometric in R Help for details. The calling syntax is

```
dgeom(x, prob)
pgeom(q, prob)
qgeom(p, prob)
rgeom(n, prob)
```

Examples

```
> # Let p = 0.1 be the probability that a molecule
> # leaves a cell at a given step
> # Then the probability that it will exit exactly
> # on the x+1 step is
> x = 0:10; p = 0.1
> dgeom(x,p)
 [1] 0.1000 0.0900 0.0810 0.0729 0.0656 0.0590 0.0531
     0.0478 0.0430 0.0387 0.0349
> # The cumulative probability that it will leave
> # before or on the x+1 step:
> pgeom(x,p)
 [1] 0.100 0.190 0.271 0.344 0.410 0.469 0.522 0.570
     0.613 0.651 0.686
> # The step by which there is at least a 50% chance
> #  that it will have exited:
> qgeom(.5,p)
 [1] 6
> # Random numbers drawn from geometric distribution
> rgeom(10,p)
 [1]   7 10 11   7   1   2 10 20   7   0
```

A conceptually related distribution is the negative bionomial distribution, which according to help(NegBinomial) "represents the number of failures that occur in a sequence of Bernoulli trials before a target number of successes is reached."

11.4.7 Exponential distribution

The exponential distribution is the continuous version of the geometric distribution. It gives the probability that the first occurrence of an event is within dt at t. It can be used to simulate the probability of radioactive decay, a unimolecular conversion, and similar processes. The distribution function for the probability that the system will have its first event at time t is $\exp(-\lambda t)$, where λ is the rate. See ?Exponential in R Help for details. The calling syntax is

```
dexp(x, rate)    # The default value of rate = 1
pexp(q, rate)
qexp(p, rate)
rexp(n, rate)
```

Examples

```
> x = seq(0,5,.2)
> plot(x,dexp(x,1), type="l")
> plot(x,pexp(x,1), type="l")
```

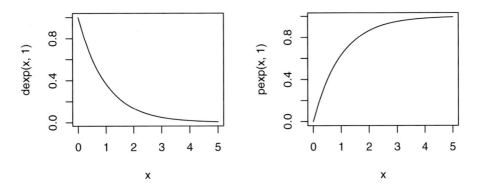

Fig. 11.15 (left) dexp and (right) pexp for exponential distribution with rate = 1

```
> qexp(c(.25,.5,.75),1)  # Quartiles
[1] 0.288 0.693 1.386

> rexp(10,1)
 [1] 0.3002 3.0558 3.2573 0.4119 1.3953 1.7069 0.1097
     0.6806 0.1546 0.0997
```

11.4.8 Other distributions

R has several other built-in distributions that may be useful on occasion. These include lognormal (`Lognormal`) and the sampling distributions Student's t (`TDist`), F (`FDist`), and Chi-squared (`Chisquare`). You should look these up in R Help.

11.5 Population Genetics

Population genetics provides excellent examples of how probability concepts can be used in biology. In the chapter on population dynamics, we assumed populations were homogeneous, while noting that age and sex were important for more detailed description. Here we deal with another important source of population heterogeneity: genetic variation. Our treatment draws heavily on that of Keen and Spain [39, Ch. 10].

11.5.1 Hardy-Weinberg Principle

The Hardy-Weinberg (HW) principle of classical genetics is based on the assumption that if mating is random in a large population of diploid individuals, matings between genotypes will occur in proportion to the fractions of those genotypes in the population. The HW principle has three major consequences:

- Gene frequencies will not change from generation to generation.
- After one generation, genotype frequencies will not change.
- Frequencies of genes and genotypes are related by a simple expression.

To illustrate, we consider a single locus with two alleles, A (dominant) and a (recessive). In a population of diploids, individuals may be AA (homozygous dominant), Aa (heterozygous), or aa (homozygous recessive). The frequencies (probabilities) of the three genotypes are denoted D, H, R, with $D+H+R=1$.

The frequency of the A gene is conventionally denoted p, while that of the a gene is $q = 1 - p$. With D, H, R known, we find

$$p = 1 - q = \frac{D+0.5H}{D+H+R} \tag{11.20}$$

$$q = 1 - p = \frac{0.5H+R}{D+H+R} \tag{11.21}$$

When a population obeying the HW assumptions reproduces, the genotype proportions in the next generation are calculated from

$$D = p^2 = (1-q)^2 \tag{11.22}$$
$$H = 2pq = 2p(1-p) = 2q(1-q) \tag{11.23}$$
$$R = q^2 = (1-p)^2 \tag{11.24}$$

If the gene frequencies (p,q) in the new generation are calculated with these values of (D,H,R) and the old (p,q), they will be found to be the same. Thus gene frequencies under HW assumptions will not change from one generation to the next. And because the gene frequencies do not change, neither will the genotype frequencies after the founding generation. The simple relation between gene and genotype frequencies is

$$(p+q)^2 = p^2 + 2pq + q^2 = D + H + R \tag{11.25}$$

In the following sections of this chapter, we examine various processes by which gene and genotype frequencies are changed.

11.5.2 Effect of selection

Under selection, different genotypes reproduce at different rates. If the most effective genotype has "fitness" of 1, less "fit" genotypes have fitness of $w < 1$. In our single locus model, we will assume that aa is less fit, so if in the first generation it has frequency R, in the next generation it has frequency wR. The frequency of the a gene will decrease accordingly:

$$q = 1 - p = \frac{0.5H + wR}{D + H + wR} \tag{11.26}$$

Then since $p = 1 - q$, the three genotype frequencies (D, H, R) in the next generation can be calculated. (See Problem 9.)

11.5.3 Selection for heterozygotes

Under some circumstances, the heterozygote is selectively favored. This is sometimes called hybrid vigor or heterosis. An example is sickle-cell anemia in native African populations, in which AA will succumb to malaria and aa to sickle-cell disease, but Aa is more resistant to malaria and suffers less severe sickle-cell disease. If the relative fitnesses of AA and aa are w_1 and w_2, and the "before" genotype frequencies are (D, H, R), the "after" gene frequency is

$$q = 1 - p = \frac{0.5H + w_2 R}{w_1 D + H + w_2 R} \tag{11.27}$$

11.5.4 Mutation

Assume reversible point mutation from A to a and back again, with forward and reverse rate constants u and v:

$$A \underset{v}{\overset{u}{\rightleftharpoons}} a \qquad (11.28)$$

At equilibrium the forward and reverse rates are equal, so that

$$up = vq = u(1-q) \Rightarrow q = \frac{u}{u+v} \qquad (11.29)$$

This equation enables us to calculate the change in frequency of the a allele due to mutation, from one generation to the next:

$$\Delta q = -vq + up = -vq + u(1-q) = -(u+v)q + u \qquad (11.30)$$

Since selection usually operates against mutations, it is useful to combine the models for selection and mutation. Given (D,H,R) in the current generation, q in the next generation will be

$$q = \frac{0.5H + wR}{D + H + wR}(1 - u - v) + u \qquad (11.31)$$

Selection and mutation will combine to give an equilibrium gene frequency. If the homozygous recessive genotype has $w = 0$, then the equilibrium frequency of a will be u, since it is maintained in the population only by mutation from A.

11.5.5 Selection involving sex-linked recessive genes

Males of many species, including humans, have only a single X chromosome, so they are essentially haploid for genes carried on that chromosome. Traits carried on the X chromosome are called sex-linked.

Consider a characteristic determined by a single pair of sex-linked alleles, A and a. The females will have genotypes (AA, Aa, aa) with frequencies (D,H,R). The frequency of a among the females is given by the usual expression

$$q_f = \frac{0.5H + R}{D + H + R} \qquad (11.32)$$

On the other hand, the frequency of allele a among males, q_m, simply equals the frequency of the a genotype. Because a male receives the X chromosome only from his mother, q_m in the current generation equals q_f of the previous generation.

We will assume that both the aa genotype in females and the a genotype in males are reduced in fitness by w relative to the AA, Aa, or A phenotypes. The female and male gene frequencies of a in the $G+1$ generation are then calculated from the frequencies in the G generation by

$$(q_f)_{G+1} = \frac{0.5H + wR}{D + H + wR} \tag{11.33}$$

$$(q_m)_{G+1} = \frac{wq_f}{wq_f + (1 - q_f)} \tag{11.34}$$

Then to update the genotype frequencies in generation $G+1$ we use the updated gene frequencies $q_m \to (q_m)_{G+1}$ and $q_f \to (q_f)_{G+1}$. Then for the females in $G+1$

$$D = (1 - q_m)(1 - q_f) \tag{11.35}$$
$$H = q_f(1 - q_m) + q_m(1 - q_f) \tag{11.36}$$
$$R = q_m q_f \tag{11.37}$$

For the males, the updated A and a genotype frequencies are $(1 - q_m)$ and q_m.

11.6 Problems

1. Verify the statement, made after eq. 11.11, "that if, within such a group, there is a 40% chance of having the disease, the probability that a positive test indicates the presence of the disease is 0.93."
2. Write an R program that will run the stochastic modeling of population growth routine 100 times and return the number of runs in which extinction occurred within 50 generations. Use the parameters given in the example: starting population of 10, average growth rate of 10%, sd of 20%.
3. In many organisms, mutations that change an AT pair to a GC pair in DNA are more common than those that change a GC pair to an AT pair. This can be expressed by a Markov chain. Write down the four conditional probabilities that a GC or AT pair at time $t+1$ was a GC or AT pair at time t, if the probability of an AT-to-GC switch is 0.002, and that of a GC-to-AT switch is 0.001. What is the probability p_{t+1} of a GC pair at $t+1$, if its probability at t was p_t? (From [1, p. 530].)
4. Use R to produce a graph of molecules inside vs. time that overlays the stochastic and analytical solutions to the Markov chain model of diffusion out of and into a cell. Use the parameters `n=50; b.tot=100; b0.i=100; p.i=0.2; p.o=0`. Plot both solutions using lines, distinguished by different colors. Add a horizontal line at the expected equilibrium value, and provide suitable title, axis labels, and legend.
5. Calculate the probability for each of the following events: (a) A standard (mean=0, sd = 1) normally distributed variable is larger than 3. (b) A normally distributed variable with mean 35 and standard deviation 6 is larger than 42. (c) Getting 10 out of 10 successes in a binomial distribution with probability 0.8. (d) $X < 0.9$ when X has the standard uniform distribution. (From [15, p. 55].)

6. Look up the `Chisquare` distribution in R Help. What is the probability that $X > 6.5$ in a `Chisquare` distribution with 2 degrees of freedom?

7. The term "six sigma" refers to an attempt to reduce errors to the point that the chance of their happening is less than the area more than six standard deviations from the mean, assuming a normal distribution. What is this area?

8. A classic problem in the history of probability theory is whether it is more likely to roll two dice 24 times and get at least one double-sixes, or to roll one die four times and get at least one six. Use the binomial distribution to decide this question.

9. Write and run an R program for simulating selection based on the HW model. Start with $(D, H, R) = (0.65, 0.27, 0.08)$ and plot q and R from 0 to 20 generations. Check that your results give constant q and stable genotype frequencies. Then alter the fitness of R by setting $w = 0.16$, and run the simulation from 0 to 80 generations. As Keen and Spain [39, p. 161] state, "This simulation is a valuable demonstration of the limited value of a eugenics program directed at a recessive gene, when selection can be exerted only on homozygous recessive individuals expressing the phenotype. When its frequency is small, the rate of removal of a recessive gene is very low."

10. Write and run a program to simulate selection for heterozygotes. Use $w_1 = 0.55$ and $w_2 = 0.15$, start with $(D, H, R) = (0.7, 0.25, 0.05)$, and graph q for 20 generations. Then demonstrate the independence of the equilibrium gene frequency from the starting frequency, using $(D, H, R) = (0.05, 0.01, 0.94)$.

11. Write an R program to simulate the combined effects of mutation and selection against a recessive allele, according to eq. 11.31. Use the parameters $u = 5 \times 10^{-5}$, $v = 1 \times 10^{-7}$, $w = 0.5$. Begin with $(D, H, R) = (0.9947, 0.005, 0.0003)$ and plot q as a function of generation until equilibrium is reached (which may take several hundred generations). (Adapted from [39, Ch. 10].)

12. Write an R program using the equations in Section 11.5.5 to simulate the effect of selection on sex-linked recessive genes. Use the parameters $w = 0.3$ and initial $q_m = q_f = 0.2$. Plot frequencies of recessive male and female phenotypes (q_m and R) for generations 0 to 8. (Adapted from [39, Ch. 10].)

Chapter 12
DNA Sequence Analysis

In this chapter we introduce some of the elementary concepts for analyzing DNA sequences in terms of "words" of length 1 (bases), 2 (base pairs), 3 (triplets, such as codons), restriction sites, etc. The analysis uses some of the basic probability concepts that we have worked with in the previous chapter. Our treatment follows closely that of Deonier et al. [18, Chs. 2 and 3].

12.1 Getting a sequence from the Web

12.1.1 Using Entrez

To analyze a sequence, we must determine it experimentally, use one that some-one else has determined, or generate it by a random simulation. Bioinformatics databases on the Web are a rich source of sequences, so we shall begin by show-ing how to access them through Entrez, "The Life Sciences Search Engine". Its home page is a cross-database search page that enables searching of many health sciences databases at the National Center for Biotechnology Information (NCBI) website. NCBI is part of the National Library of Medicine (NLM), which is itself a department of the National Institutes of Health (NIH).

To access Entrez, go to http://www.ncbi.nlm.nih.gov/gquery/gquery.fcgi (or sim-ply Google "Entrez"). Suppose we are interested in the complete genome sequence of the bacterium *E. coli*. From the Entrez home page, click Genome. Choose Bac-teria Chromosome in the left-hand column, scroll down to Escherichia coli str. K12 substr. MG1655, and click on NC000913 in the second column. In the resulting ta-ble, click GenBank: U00096 in the Genome Info column. Set the display to FASTA, Show 1, Send to Text, Range from 1 to 1000 (or whatever you wish), and click Refresh. The result is shown on p. 235

One can ask many questions when given such a sequence, about its function, how it differs from other organisms or from different parts of the genome of the same

V. Bloomfield, *Computer Simulation and Data Analysis in Molecular Biology and Biophysics,*   233
Biological and Medical Physics, Biomedical Engineering, DOI: 10.1007/978-1-4419-0083-8_12,

organism, how it might have evolved, etc. This is the province of bioinformatics, about which in this chapter we will touch only on some very elementary aspects. We do so by considering how to analyze the statistics of some very short sequences: mononucleotides, dinucleotides, and restriction sequences.

12.1.2 Making the sequence usable by R

To proceed further, we need to turn this 1000-character string into a vector that R can process. Fortunately, R has the necessary functions. (See the Appendix for a summary of basic text manipulation functions in R.) The function strsplit(x, split) splits a string x into characters (or groups of characters) at each occurrence of the substring split. strsplit returns a list, which we convert into a character vector with the function unlist.

If we apply this sequence of operations to the first two lines (denoted by .12 in the name) of the *E. coli* sequence above, using the null string " " as split, we get the result in the top block on p. 236.

All looks well, except for the "\n" (newline) characters in positions 71 and 142. These can be eliminated by redefining K12.12.vec to include only those elements that are not "\n", with results shown in the bottom block on p. 236.

12.1.3 Converting letters to numbers

For some purposes (e.g., to plot a property of the bases vs. sequence) it may be desirable to convert the letters to numbers. To do so, one can use the function gsub(x, y, z), which replaces pattern x with pattern y in character vector z. As an example, the following code shows the results of replacing A by 1 in the first 10 positions of K12.12.vec.

```
> gsub("A","1",K12.12.vec)[1:10]
 [1] "1" "G" "C" "T" "T" "T" "T" "C" "1" "T"
```

To replace A by 1, C by 2, G by 3, and T by 4, we can apply gsub() recursively and then convert the number characters to numbers with as.numeric:

```
> K12.12.numchar = gsub("T","4",gsub("G","3",
    gsub("C","2",gsub("A","1",K12.12.vec))))
> K12.12.num = as.numeric(K12.12.numchar)
> K12.12.num[1:10]
 [1] 1 3 2 4 4 4 4 2 1 4
```

```
>gi|48994873:1-1000 Escherichia coli str. K-12 substr. MG1655,
    complete genome
AGCTTTTCATTCTGACTGCAACGGGCAATATGTCTCTGTGTGGATTAAAAAAAGAGTGTCTGATAGCAGC
TTCTGAACTGGTTACCTGCCGTGAGTAAATTAAAATTTATTGACTTAGGTCACTAAATACTTTAACCAA
TATAGGCATAGCGCACAGACAGATAAAAATTACAGAGTACACAACATCCATGAAACGCATTAGCACCACC
ATTACCACCACCATCACCATTACCACAGGTAACGGTGCGGGCTGACGCGTACAGGAAACACAGAAAAAAG
CCCGCACCTGACAGTGCGGGCTTTTTTTTTCGACCAAAGGTAACGAGTAACAACCATGCGAGTGTTGAA
GTTCGGCGGTACATCAGTGGCAAATGCAGAACGTTTTCTGCGTGTTGCCGATATTCTGGAAAGCAATGCC
AGGCAGGGGCAGGTGGCCACCGTCCTCTCTGCCCCCGCCAAAATCACCAACCACCTGGTGGCGATGATTG
AAAAAACCATTAGCGGCCAGGATGCTTTACCCAATATCAGCGATGCCGAACGTATTTTTGCCGAACTTTT
GACGGGACTCGCCGCGCCAGCCGGGGTTCCCGCTGGCCAATTGAAAACTTTCGTCGATCAGGAATTT
GCCCAAATAAAACATGTCCTGCATGGCATTAGTTTGTTGGGGCAGTGCCGGCATCAACGCTGCGC
TGATTTGCCGTGGCGAGAAAATGTCGATCGCCATTAGTTGTTGGGGCAGTGCCGGCATCAACGCTGCGC
TACTGTTATCGATCCGGTCGAAAAACTGCTGGCAGTGGGCATTACCTCGAATCTACCGTCGATATTGCT
GAGTCCACCCGCCGTATTGCGGCAAGCCCATTCCGGCTGATCACATGGTGCTGATGCGGCAGGTTTCACCG
CCGGTAATGAAAAGGCGAACTGGTGTGCTTGGACGCAACGGTTCCGACTACTCTGCTGCGGTGCTGGC
TGCCTGTTTACGCGCCGATT
```

```
> K12.12 = "AGCTTTTCATTCTGACTGCAACGGGCAATATGTCTCTGTGTGGATTAAAAAAGAGTGTCTGATAGCAGC
TTCTGAACTGGTTACCTGCCGTGAGTAAATTAAAATTTATTGACTTAGGTCACTAAATACTTTAACCAA"
> K12.12.vec = unlist(strsplit(K12.12,""))
> K12.12.vec
  [1] "A" "G" "C" "T" "T" "T" "T" "C" "A" "T" "T" "C" "T"
 [14] "G" "A" "C" "T" "G" "C" "A" "A" "C" "G" "G" "G" "C"
 [27] "A" "A" "T" "A" "T" "G" "T" "C" "T" "C" "T" "G" "T"
 [40] "G" "T" "G" "G" "A" "T" "T" "A" "A" "A" "A" "A" "A"
 [53] "G" "A" "G" "T" "G" "T" "C" "T" "G" "A" "T" "A" "G"
 [66] "C" "A" "G" "C" "\n" "T" "T" "C" "T" "G" "A" "A" "C"
 [79] "T" "G" "G" "T" "T" "A" "C" "C" "T" "G" "C" "C" "G"
 [92] "T" "G" "A" "G" "T" "A" "A" "A" "T" "T" "A" "A" "A"
[105] "A" "T" "T" "T" "A" "T" "T" "G" "A" "C" "T" "T" "A"
[118] "G" "G" "T" "C" "A" "C" "T" "A" "A" "A" "T" "A" "C"
[131] "T" "T" "T" "A" "A" "C" "C" "A" "A" "\n"

> K12.12.vec = K12.12.vec[K12.12.vec != "\n"]
> K12.12.vec
  [1] "A" "G" "C" "T" "T" "T" "T" "C" "A" "T" "T" "C" "T" "G" "A" "C"
 [17] "T" "G" "C" "A" "A" "C" "G" "G" "G" "C" "A" "A" "T" "A" "T" "G"
 [33] "T" "C" "T" "C" "T" "G" "T" "G" "T" "G" "G" "A" "T" "T" "A" "A"
 [49] "A" "A" "A" "A" "G" "A" "G" "T" "G" "T" "C" "T" "G" "A" "T" "A"
 [65] "G" "C" "A" "G" "C" "T" "T" "C" "T" "G" "A" "A" "C" "T" "G" "G"
 [81] "T" "T" "A" "C" "C" "T" "G" "C" "C" "G" "T" "G" "A" "G" "T" "A"
 [97] "A" "A" "T" "T" "A" "A" "A" "A" "T" "T" "T" "A" "T" "T" "G" "A"
[113] "C" "T" "T" "A" "G" "G" "T" "C" "A" "C" "T" "A" "A" "A" "T" "A"
[129] "C" "T" "T" "T" "A" "A" "C" "C" "A" "A"
```

12.2 Single base sequences and frequencies

12.2.1 Counting the number of each type of base

For duplex DNA, we know that G = C and A = T, so the fraction of Cs, for example, equals fr(G+C)/2. Also, since fr(A+T) = 1 − fr(G+C), a single number such as fr(G+C) specifies the proportion of all four nucleotides in the duplex. However, this is not true of a single strand, since there may be compositional biases on one strand that are complemented by the other strand. Most commonly, the sequence given in the database is that of the leading strand, whose 5'-to-3' direction is the same as that in which the replication fork moves.

To calculate the total number of bases, use the `length` function on the vector of bases:

```
> length(K12.12.vec)
[1] 140
```

To calculate the number of bases of each type, we can use the logical code

```
> length(K12.12.vec[K12.12.vec == "A"])
[1] 43
```

12.2.2 Simulating a random sequence of bases

It is often useful to compare an observed DNA sequence against one generated at random, to get an idea of how the real sequence deviates from randomness (which means that it contains information). This piece of R code generates 10,000 samples with replacement from the vector of base symbols `base` assuming equal probabilities $1/4 = 0.25$ for each of the four bases. To keep from printing out 10,000 symbols, we put the result of sample in an intermediate variable `sequence`, and then print just the first 15 values of `sequence`.

```
> p = c(0.25,0.25,0.25,0.25)
> base = c("A","C","G","T")
> sequence = sample(base,10000,replace=TRUE,p)
> sequence[1:15]
 [1] "G" "T" "A" "C" "T" "G" "T" "T" "A" "C" "G" "T"
     "G" "T" "T"
```

Another way to generate a random sequence, in which the letter representations of the bases are printed as a single string, comes from the R & BioCon Manual:

```
> x = as.integer(runif(20, min=1, max=5))
> x[x==1] = "A"; x[x==2] = "T"; x[x==3] = "G"
> x[x==4] = "C"
```

```
> paste(x, sep = "", collapse ="")
[1] "GATTGCGGCCTTCAATCTCG"
```

In this code, as.integer() rounds down the values of the uniform random numbers, giving integers between 1 and 4. The code x[x==1] = "A" takes those
elements which are identically equal to 1 (hence the ==) and sets them equal to
A, and so on for the other three bases. paste() is the R function that concatenates strings, with sep and collapse as separators (here zero spaces) between
the string elements. See R Help for more information.

The model we have used, in which the probability of a given base being found in
a given position is identical and independent of position, is often called the "independent and identically distributed" or "iid" model.

12.2.3 Simulating the distribution of the number of A residues in a sequence

The preceding example gave just one realization of a potential random sequence.
Each generation of a random sample will lead to somewhat different results. Here
we follow Deonier et al. [18, p. 47] in generating 2000 samples of a 1000-base
length of DNA, and asking how the number of As is distributed in the 2000 samples.
We use the binomial distribution, with the probability of finding an A = 0.25. (The
probability of finding not-A, i.e., C, G, or T, is 0.75). We expect the mean to be
$(1/4) \times 1000 = 250$, and the result is very close, though from the histogram we see
that in the 2000 samples there are as few as about 210 As and as many as about
290. The expected standard deviation is $[np(1-p)]^{1/2} = (1000 \times 0.25 \times 0.75)^{1/2} =$
13.69.

```
> options(digits=4)
> x = rbinom(2000,1000,0.25)
> mean(x)
[1] 249.2
> sd(x)
[1] 13.84
> hist(x,xlab="Number of As", main="")
```

The probability of getting at least 280 A's is

$$P(n_A \geq 280) = \sum_{k=280}^{1000} \frac{1000!}{k!(1000-k)!} p_A^k (1-p_A)^{1000-k} = 1 - \text{pbinom}(279, 1000, p_A)$$

(12.1)

which evaluates to

```
> p = .25; n = 1000
> 1-pbinom(279,n,p)
[1] 0.01644
```

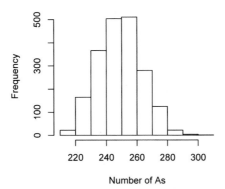

Fig. 12.1 Distribution of the number of A residues in 2000 samples of a 1000-base length of DNA, assuming $p_A = 0.25$

12.3 Dinucleotide sequences and frequencies

Dinucleotides are important, at least in part, because their sequence influences the bending and twisting of the DNA double helix, which may in turn influence gene expression.

12.3.1 Probability of observing dinucleotides

For the iid model, the probability that base i is followed by base j is simply the product of observing the individual bases, since the probabilities are independent and thus multiply:

$$Pr(j|i) = Pr(i)Pr(j) = p_i p_j \tag{12.2}$$

If there are n bases in the sequence, there are $(n-1)$ dinucleotides, so the expected number of dinucleotides (i, j) is $E_{ij} = (n-1)p_i p_j$. If the observed number is O_{ij}, then a standard measure of whether the difference between observed and expected is beyond random fluctuations is the χ^2 (Chi-squared) statistic:

$$\chi^2 = \frac{(O-E)^2}{E} \tag{12.3}$$

Large values of χ^2 suggest that the iid model is a poor one. Deonier et al. [18, p. 49] give a recipe for calculating whether χ^2 is unlikely. First, calculate the number c given by

$$c = \begin{cases} 1 + 2p_i - 3p_i^2, & i = j \\ 1 - 3p_i p_j, & i \neq j \end{cases} \tag{12.4}$$

then calculate χ^2/c. If this ratio is greater than 3.84, we can conclude that the iid model is not a good one.

Table 12.1, for example, is a table from Deonier et al. [18, p. 50] that shows χ^2/c for the first 1000 nucleotides of two microbes. For *E. coli*, the base frequencies were taken as 0.25 for each. For *M. genitalium*, p(A,C,G,T) were taken as (0.45, 0.09, 0.09, 0.37). Significantly non-random values are in boldface.

Table 12.1 Observed χ^2/c for

Dinucleotide	*E. coli*	*M. genitalium*
AA	**6.78**	0.15
AC	0.05	1.20
AG	**5.99**	0.18
AT	0.01	0.01
CA	2.64	0.01
CC	0.03	0.39
CG	0.85	**4.70**
CT	**4.70**	1.10
GA	2.15	0.34
GC	**10.04**	1.07
GG	0.01	0.09
GT	1.76	0.61
TA	**5.99**	1.93
TC	**9.06**	2.28
TG	3.63	0.05
TT	1.12	0.13

We see that *E. coli* does not fit the iid model very well, while *M. genitalium* deviations are not so large.

12.3.2 Markov chain simulation of dinucleotide frequencies

In the previous chapter we introduced Markov chains for situations when the outcome of event $i+1$ is influenced by the immediately preceding event i. The application to dinucleotide frequencies should be apparent, but the mathematical apparatus needs some amplification since we have four events, each of which can be influenced by the four possible preceding events. We handle this using matrix notation. (Note: In this section we use some matrix algebra. If you need a review, there are many on-line sources as well as textbooks.) We introduce a transition matrix P:

$$P = \begin{pmatrix} p_{AA} & p_{AC} & p_{AG} & p_{AT} \\ p_{CA} & p_{CC} & p_{CG} & p_{CT} \\ p_{GA} & p_{GC} & p_{GG} & p_{GT} \\ p_{TA} & p_{TC} & p_{TG} & p_{TT} \end{pmatrix} \tag{12.5}$$

in which each row corresponds to the possible states in position i, and each column corresponds to the possible states in position $i+1$. For example, $P_{12} = p_{AC}$ corresponds to the probability that an A at position 1 will be followed by a C at position 2. We assume that the Markov chain is homogeneous, so that these probabilities do not depend on the values of i, but are the same throughout the sequence. Since position i must be followed by one of the four bases in position $i+1$, the row probabilities must sum to unity: $\sum_j P_{ij} = 1$.

We follow Deonier et al [18, pp. 54–55] in constructing the Markov transition matrix for the bacterium *M. genitalium*. Begin with the matrix dnf of observed dinucleotide frequencies:

```
> # Enter the frequencies, and name rows and columns
> dnf = matrix(c(0.146,0.052,0.058,0.089,0.063,0.029,
  0.010,0.056,0.050,0.030,0.028,0.051,0.087,0.047,0.063,
  0.140),4,4,byrow=T,
  dimnames=list(c("A","C","G","T"),c("A","C","G","T")))
> # Display the result
> dnf
      A     C     G     T
A 0.146 0.052 0.058 0.089
C 0.063 0.029 0.010 0.056
G 0.050 0.030 0.028 0.051
T 0.087 0.047 0.063 0.140
```

From this we calculate the transition matrix P using the total probability requirement by taking each element of dnf and dividing by the sum of elements in that row:

```
> # First dimension and name the matrix P
> P = matrix(rep(0,16),nrow=4,
  dimnames=list(c("A","C","G","T"),c("A","C","G","T")))
> # Then calculate the elements
> for (i in 1:4) {
for (j in 1:4) {
P[i,j] = dnf[i,j]/sum(dnf[i,])}
}
> # Display the result, with options(digits=3)
> P
      A     C     G     T
A 0.423 0.151 0.168 0.258
C 0.399 0.184 0.063 0.354
G 0.314 0.189 0.176 0.321
T 0.258 0.139 0.187 0.415
```

To specify the evolution of the Markov chain, we need to know its initial state. Since this is a probability calculation, the initial state may be specified by an initial probability distribution such as the fractions of the four bases at the beginning of the chain. We represent this as a row vector $\pi = (p_A, p_C, p_G, p_T)$, where the four components are the mole fractions of (A,C,G,T). The law of total probability then says that

$$Pr(X_2 = j) = \sum_{i=1}^{4} Pr(X_2 = j, X_1 = i) = \sum_{i=1}^{4} Pr(X_1 = i)Pr(X_2 = j|X_1 = i) = \sum_{i=1}^{4} \pi_i P_{ij}$$

$$(12.6)$$

The last in this chain of equalities is the product of the row vector π with the matrix P, so the probability of the state of the second position in the chain can be written in matrix notation as

$$\pi^{(2)} = \pi P \qquad (12.7)$$

Repetition of this line of argument leads to

$$\pi^{(3)} = \pi^{(2)} P = \pi P^2 \qquad (12.8)$$

and so on. Calculation of the matrix power can be readily done with the function

```
> matpower = function (M, n) {
Mp = M
if (n == 1) Mp = M else
for (i in 1:(n-1)) Mp = Mp %*% M
return (Mp)
}
```

Remember that the operator for matrix multiplication, or for multiplication of a vector and a matrix, is %*%.

The double-stranded *M. genitalium* genome is 31.6% G+C, so if we start out with π as the mole fractions of the four bases, we find that the process above converges to the limiting distribution after only two steps.

```
> pi = c(0.342,0.158,0.158,0.342)
> pi%*%matpower(P,1)
            A       C       G       T
[1,]  0.346  0.158  0.159  0.337
> pi%*%matpower(P,2)
            A       C       G       T
[1,]  0.346  0.158  0.159  0.336
> pi%*%matpower(P,3)
            A       C       G       T
[1,]  0.347  0.158  0.159  0.336
```

Even when starting with equal mole fractions of the four bases, the limiting distribution is reached after three steps.

We can now use this Markov process to simulate a sequence that resembles that of *M. genitalium* not just in base composition, but also in dinucleotide frequencies. (See [18, pp. 55–57].) We define a function markovseq that will generate a sequence of desired length n, given the matrix P of transition probabilities, the vector x of base possibilities, and the limiting distribution π of base mole fractions, which gives the initial sampling probability of picking one of the bases. At each step, we use `sample` to pick the next base according to the relevant probability vector `pi` or `mseq[i,]`.

```
> markovseq = function(x,pi,P,n){
# Initialize sequence vector of length n
mseq = rep(0,n)
# Choose initial element
mseq[1] = sample(x,1,replace=TRUE,pi)
# Choose subsequent elements using P
for (i in 1:(n-1)){
mseq[i+1] = sample(x,1,replace=T,P[mseq[i],])
}
return(mseq)
}
```

We use this function to generate one of the many possible sequences of length 50,000, using the data for *M. genitalium*, and check that it has approximately the correct proportion of A residues.

```
> x = c("A","C","G","T")
> mgenseq = markovseq(x,pi,P,50000)
> length(mgenseq[mgenseq=="A"])/50000
[1] 0.343
```

This is close to the 34.2% from the initial data. To calculate the fraction of AA dinucleotides, for example, we use the following code.

```
> nAA = 0 # Initialize count
> for (i in 1:49999) {
if (mgenseq[i]=="A" && mgenseq[i+1]=="A") nAA = nAA + 1
}
> nAA
[1] 7292
> nAA/49999
[1] 0.146
```

This result matches exactly the AA entry in the dinucleotide frequency matrix `dnf`. Of course, since this is a random process, other runs will give slightly different values.

One can go on to "words" of length = 3, and investigate such things as codon utilization frequencies [18, pp. 57–60]. We shall not deal with this, but will move to even longer sequences as applied to restriction site distributions.

12.4 Simulation of restriction sites

In a random or pseudo-random DNA sequence, the occurrence of restriction sites should roughly follow a Poisson distribution. According to a Poisson model, the distribution of restriction fragment lengths should be approximately exponential, and this is often seen in real sequences. In this section we generate a sequence according to the iid model, and compute the fragment sizes. The expected mean number of sites for an m-base restriction enzyme acting on a DNA molecule of length n in which all of the bases have equal probability of $1/4$ is $n(1/4)^m$. For the example of lambda bacteriophage DNA, with approximately 48,500 base pairs, and the 4-base *Alu* restriction enzyme that cleaves at AGCT sequences, this works out to 190. Our treatment follows that of Deonier et al. [18, Ch. 3, pp. 85-89].

We first generate a DNA sequence 48,500 bases long, using the iid model with all base probabilities = 0.25. As before, the sample function does yeoman duty.

```
> x = c("A","C","G","T")
> p = c(0.25, 0.25, 0.25, 0.25)
> n = 48500
> lambdaseq = sample(x,n,replace=T,p)
> lambdaseq[1:10] # Check that it ran
[1] "T" "G" "C" "A" "G" "C" "C" "A" "G" "G"
> length(lambdaseq) # Check that the length is correct
[1] 48500
```

Now we define a function that matches the restriction sites found in a DNA sequence DNAseq with the enzyme recognition sequence rseq, and tabulates their start positions.

```
> restrictionsites = function(DNAseq, rseq) {
# Length of DNA sequence
lDNAseq = length(DNAseq)
# Initialize vector to hold site positions
DNApos = rep(0,lDNAseq)
# Length of restriction site sequence
m = length(rseq)
# Records whether position of DNAseq matches rseq
match = rep(0,m)
# Check each position in DNAseq to see if a site
  starts there
for (i in 1:(lDNAseq-m+1)) {
  for (j in 1:m){
    # Match at position j
    if (DNAseq[i+j-1] == rseq[j]) match[j] = 1
    }
  # Record start position if all positions match
  if (sum(match) == m) DNApos[i] = i
```

```
  # Reinitialize for next iteration of loop
  match = rep(0,m)
  }
# Vector of start site positions
rpos = DNApos[DNApos > 0]
return(rpos)
}
```

Now we use `restrictionsites` to find the *Alu* sites in the random `lambdaseq` we have generated. The nested loops may cause a noticeable delay before the function has done its work.

```
> alu = c("A","G","C","T")
> alu.lambda.map = restrictionsites(lambdaseq, alu)
> # Check the number of sites identified
> length(alu.lambda.map)
[1] 159
> # Positions of the first ten sites
> alu.lambda.map[1:10]
 [1]    23    63   143   256   350   419   688  1516  1872  1900
```

With this particular random sequence, we found 199 *Alu* restriction sites, a bit more but in the range of the 189 predicted from Poissson statistics. More runs with different sequences would be necessary to get a mean value of the number of sites.

To compare a restriction digest with experimental data such as gel electrophoresis, we need to know the fragment lengths, i.e., the distance between restriction sites. The function `fraglen` gives us those lengths. It takes into account that we are dealing with a linear DNA molecule, and so will have fragments bounded by the beginning and end of the molecule.

```
> fraglen = function(resmap, n) {
# resmap is the vector of restriction sites
# in a linear DNA molecule
# n is the length (base pairs) of the DNA molecule
# Initialize vector for fragment lengths
fragments = rep(0, length(resmap))
# Augment resmap with inclusion of end piece
resmap = c(resmap,n)
# Subtract successive start sites to get lengths:
for (i in 1:(length(resmap)-1))
  fragments[i] = resmap[i+1] - resmap[i]
  # Augment with left end piece
  fragments = c(resmap[1], fragments)
return(fragments)
}
```

Applying this function to our simulated restriction digest of lambda DNA by *Alu*, we find

```
> alu.lambda.fragments = fraglen(alu.lambda.map,48500)
> alu.lambda.fragments[1:10]
[1]   23   40   80  113   94   69 269 828 356   28
> max(alu.lambda.fragments)
[1] 1678
> min(alu.lambda.fragments)
[1] 4
> # Number of restriction sites + 1
> length(alu.lambda.fragments)
[1] 200
# Sum to get total length of DNA molecule
> sum(alu.lambda.fragments)
[1] 48500
```

A histogram of the fragment lengths, generated by hist(alu.lambda.fragments,
roughly shows the expected exponential distribution.

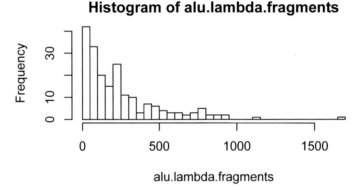

Fig. 12.2 Distribution of fragment lengths of λ DNA digested by *Alu* restriction endonuclease

12.5 Detecting periodicity in a sequence

The composition and physicochemical features of nucleic acids and proteins are
often found to repeat with some periodicity along the sequence. Such periodicity
is often hidden in an otherwise apparently random sequence, but can be detected
with appropriate mathematical tools. These tools are generally based on Fourier
transform methods.

In R, the most convenient tool for detecting periodicity is the spectrum function, which produces a plot called a periodogram. (R also has a fast Fourier
transform function, fft, but it is less convenient to use.) R Help says "The spec-

trum function estimates the spectral density of a time series." In fact, "time" can be any regularly progressing variable, such as position along a sequence, wavelength in an absorption spectrum, etc.

As an example, suppose a given DNA sequence has iid character except at every tenth base, where there is a high probability of finding an A. Suppose also that we have a way of detecting only As, which give a signal = 1, while the other bases give a signal = 0. We can generate such a random sequence of signals with the first two lines of the following code. The first line generates 1000 random numbers, with zeros appearing 3/4 of the time and ones 1/4 of the time. The `for` loop then goes back over the sequence, checks whether it is at a position evenly divisible by 10 with the integer division operator `%/%`, and if so places a 1 with probability 0.8.

```
> x = sample(c(0,1),1000,replace=T,prob=c(0.75,0.25))
> for (i in 1:1000) {
    if (i%/%10 == i/10)  x[i] = sample(c(0,1),1,
      replace=T,prob=c(0.2,0.8))
    }
```

When the resulting sequence of ones and zeros is plotted, it is virtually impossible to discern any periodicity by eye. However, applying the `spectrum` function clearly shows peaks at 1/10 and multiples of that frequency. The parameter `spans` sets the size of the window over which the spectrum is averaged. Its values may be varied to find the best smoothing.

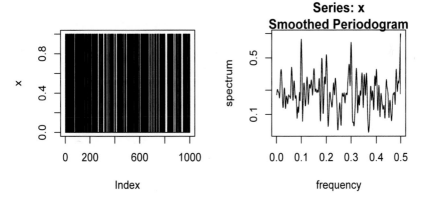

Fig. 12.3 Simulated distribution (left) and periodogram (right) of A residues in a DNA sequence

```
> par(mfrow=c(1,2))
> plot(x,type="h")
> spectrum(x,spans=c(5,5))
> par(mfrow=c(1,1))
```

12.6 Problems

1. Download the sequence of the complete genome of lambda bacteriophage (Enterobacteria phage lambda, NC 001416) from Entrez in FASTA format, and convert it to a vector of letters (A,C,G,T). Print its length, and the first 10 and the last 10 letters in the sequence, to check that the process was correct. You will use this sequence in the following problems.
2. Count the number of bases of each type, and show that they add to the total number of bases in the lambda genome.
3. Purine bases (G,A) are denoted by R, and pyrimidine bases (C,T) by Y. Convert the lambda sequence to a sequence of Rs and Ys, and count the number of each.
4. Write a function to determine the number of base pairs of each type (AA, AC, etc.) and use the function to enter the fractions of each dinucleotide type into a 4×4 matrix like dnf. Show that the fractions add to 1.
5. Use the matrix in the preceding problem to determine the Markov transition matrix for lambda DNA.
6. Find the limiting distribution of bases for lambda, using the Markov matrix you have just determined. How many steps does it take to converge?
7. Repeat the previous three problems for the purine, pyrimidine representation of the DNA sequence. Your matrices will be 2×2.
8. The recognition sequence for the restriction endonuclease EcoR1 is GAATTC. Find the locations of the EcoR1 restriction sites in lambda DNA. How many fragments will be produced, and what are their lengths?
9. Run the spectrum function on the purine, pyrimidine representation of lambda DNA to see if there are any discernible periodicities.
10. Use the observed dinucleotide frequencies for lambda DNA, along with those expected from the mole fractions of the bases, to calculate χ^2 and then χ^2/c according to the recipe in eqs. 12.2–12.4. Construct a table like Table 12.1 for lambda. Are there any dinucleotides that show improbable frequencies?

Part IV
Statistical Analysis in Molecular and Cellular Biology

Chapter 13
Statistical Analysis of Data

Molecular biologists and biophysicists have a lot of data to analyze, and R has a lot of tools to help with the analysis. We will consider three major topics: summary statistics for a single group of data, statistical comparison of two samples, and analysis of spectral data. A concise and useful reference for this chapter is P. Dalgaard's *Introductory Statistics with R* [15], especially Ch. 3 on "Descriptive statistics and graphics" and Ch. 4 on "One- and two-sample tests". The standard comprehensive book on statistics using R is that by Venables and Ripley [64]. Other useful books are those by Maindonald and Braun [43], Verzani [65], and Everitt and Hothorn [23].

13.1 Summary statistics for a single group of data

When you've made a series of measurements on a single sample, the first thing to do is to get an overview of the data: its mean, dispersion, range, etc. The statistical and graphical functions in this section can provide that overview.

13.1.1 Summary statistics

Consider a simulated set of 100 observations assembled in a vector x, normally distributed about a mean of 2 and a standard deviation of 0.5.

```
> x = rnorm(100,2,.5) # Mean of 2, sd of 0.5
> mean(x)
[1] 2.066552
> sd(x)
[1] 0.4779595
> var(x) # = sd^2
[1] 0.2284453
```

V. Bloomfield, *Computer Simulation and Data Analysis in Molecular Biology and Biophysics,*   251
Biological and Medical Physics, Biomedical Engineering, DOI: 10.1007/978-1-4419-0083-8_13,
© Springer Science + Business Media, LLC 2009

```
> median(x) # Also near 2 for symmetrical distribution
[1] 2.055512
```

Quantiles by default divide the distribution into 25% slices, which are also called quartiles. The interquartile range IQR, which gives the difference between 75% and 25% quantiles, is an alternative to the standard deviation for characterizing the breadth of a distribution.

```
> quantile(x)
      0%       25%       50%       75%      100%
1.018998 1.707025 2.055512 2.362961 3.176611
> IQR(x)
[1] 0.6559363
```

We can compute quantiles for other intervals, e.g., every 20%:

```
> pvec=seq(0,1,.2)
> quantile(x,pvec)
      0%       20%       40%       60%       80%      100%
1.018998 1.674610 1.956470 2.203860 2.438100 3.176611
```

We can get some of the desired statistics by the single command summary

```
summary(x)
   Min. 1st Qu.  Median    Mean 3rd Qu.    Max.
  1.019   1.707   2.056   2.067   2.363   3.177
```

13.1.2 Histograms

It is generally desirable to augment numerical summaries of the data with graphical presentations. Graphs can quickly reveal outliers or unusual features, and the qqnorm plot described below can indicate whether it is valid to use statistical tests that assume the data are normally distributed. To visualize x we can plot a histogram:

```
> hist(x)
```

We can set the number of bars or bins with breaks=n:

```
> hist(x,breaks=5)
```

or even specify uneven bin sizes:

```
> brk = c(0.5,1,1.5,1.8,2,2.1,2.3,2.5,3,3.5)
> hist(x,breaks=brk)
```

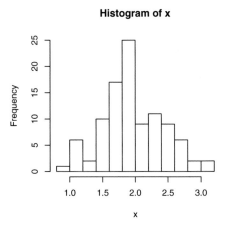

Fig. 13.1 Default histogram of 100 normally distributed random numbers with mean = 2 and sd = 0.5

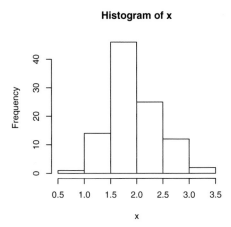

Fig. 13.2 Histogram with five breaks specified

13.1.3 Cumulative distribution

The cumulative probability of finding a particular value in x is plotted by first sorting x, using the sort function, and then plotting on the y axis a value which goes from $1/n$ to 1. The type = "s" option calls the step style of plot.

```
> n = length(x)
> plot(sort(x),(1:n)/n,type="s",ylim=c(0,1))
```

Histogram of x

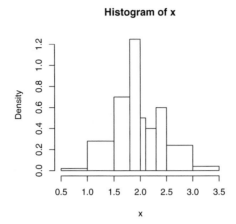

Fig. 13.3 Histogram with uneven bin sizes specified

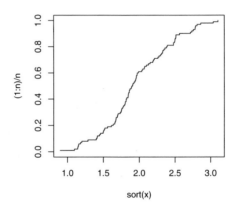

Fig. 13.4 Cumulative distribution of 100 normally distributed random numbers with mean = 2 and sd = 0.5

13.1.4 Test of normal distribution using qqnorm

Many statistical tests depend on the assumption that the variations in a parameter are normally distributed. The easiest way to see whether this assumption holds is by the qqnorm function in R, which plots the kth smallest observation against the expected value of the kth smallest observation out of n in a standard normal distribution. This should be approximately a straight line if the observations are drawn from a normal distribution with any mean and standard deviation.

```
> qqnorm(x)
```

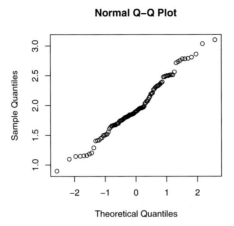

Fig. 13.5 qqnorm plot of normally distributed random numbers

This comes adequately close to a straight line except at the very ends, while a random sample drawn from another distribution (in this case a uniform random distribution) does not:

```
> xuni = runif(100)
> qqnorm(xuni)
```

If desired, the qqline command can be used to draw the best straight line through the qqnorm points.

13.1.5 Boxplots

A good way to graphically summarize a distribution of values is through a "boxplot" (better described as a box-and-whiskers plot). This plot is particularly useful for non-normal distributions, where instead of mean and standard deviation the best descriptors are median and interquartile range (IQR). As an example we use the data for IgM (a vector of Serum IgM (g/L) in 298 children aged 6 months to 6 years from Altman (1991) [4] as included in the ISwR package of Dalgaard, which may be downloaded from the CRAN repository.

```
> library(ISwR)
> data(IgM)
> par(mfrow=c(1,2))
> boxplot(IgM); title(main="IgM")
```

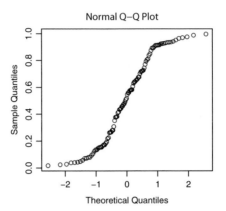

Fig. 13.6 `qqnorm` plot of uniformly distributed random numbers

```
> boxplot(log(IgM)); title(main="log(IgM)")
> par(mfrow=c(1,1))
```

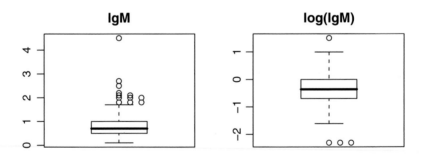

Fig. 13.7 Boxplot of IgM data

The box indicates the "hinges"—essentially the 1/4 and 3/4 quartiles; the thick line is the median. The lines (whiskers) indicate the largest and smallest observations that lie within 1.5 times the box size from the nearest hinge. Extreme points farther out are shown individually. We see that the data are approximately symmetrically distributed when expressed on a logarithmic basis.

In the above code, "mfrow" stands for "multiframe, rowwise, 1 × 2 layout". There is a corresponding mfcol command. It's necessary to issue a par(mfrow=c(1,1)) command at the end so that subsequent plots have the normal 1 × 1 arrangement.

13.1.6 Summary statistics for grouped data using tapply()

Sometimes you will have a single variable, but grouped according to values of some other factor. We work with the red.cell.folate data frame from the ISwR package, which gives folate concentrations (μg/L) in red cells after three conditions of ventilation during anesthesia. The data frame codes ventilation as "a factor with levels N2O+O2,24h: 50% nitrous oxide and 50% oxygen, continuously for 24 hours; N2O+O2,op: 50% nitrous oxide and 50% oxygen, only during operation; O2,24h: no nitrous oxide but 35%50% oxygen for 24 hours." (Data from [4].)

After loading the ISwR package and stating that we want to work with the red.cell.folate data, we "attach" the data, so we can access folate and ventilation as separate variables. We then generate boxplots of folate concentrations for the three ventilation conditions.

```
> library(ISwR)
> data(red.cell.folate)
> attach(red.cell.folate)
> boxplot(folate~ventilation)
```

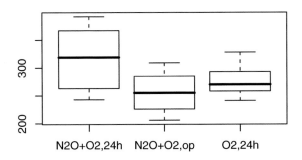

Fig. 13.8 Boxplot of red.cell.folate data for three ventilation conditions

The function `tapply` enables calculation of means, standard deviations, etc. for each group.

```
> tapply(folate,ventilation,mean)
N2O+O2,24h  N2O+O2,op        O2,24h
      316.6        256.4         278.0
> tapply(folate,ventilation,sd)
N2O+O2,24h  N2O+O2,op        O2,24h
      58.72        37.12         33.76
> tapply(folate,ventilation,length)
N2O+O2,24h  N2O+O2,op        O2,24h
          8            9             5
```

We can make a nicer-looking tabular presentation by defining the various `tapply` results as variables, and then combining them as column vectors with `cbind`, giving them more meaningful names in the process:

```
> xbar = tapply(folate,ventilation,mean)
> s = tapply(folate,ventilation,sd)
> n = tapply(folate,ventilation,length)
> cbind(mean=xbar, std.dev=s, n=n)
              mean std.dev n
N2O+O2,24h 316.6    58.72 8
N2O+O2,op  256.4    37.12 9
O2,24h     278.0    33.76 5
```

This code shows how to plot stacked histograms of the three ventilation conditions:

```
> par(mfrow=c(3,1))
> hist(folate[ventilation=="N2O+O2,24h"])
> hist(folate[ventilation=="N2O+O2,op"])
> hist(folate[ventilation=="O2,24h"])
> par(mfrow=c(1,1))
```

13.2 Statistical comparison of two samples

Inspection of biochemistry journals shows that the overwhelming usages of statistics are of two types: comparing two samples, and fitting data to a model to extract the parameters of the model. This section is devoted to statistical tools for comparing two samples (or—an equivalent problem —comparing a single sample against a predetermined standard). For the most part, we will do this using the `t.test` function.

Wikipedia has an informative entry on the t test. The historical note is interesting:

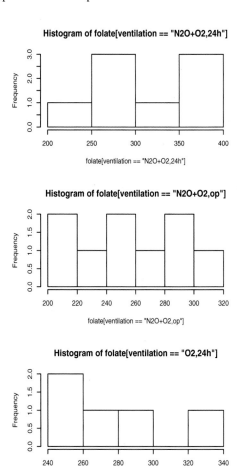

Fig. 13.9 Stacked histograms of `red.cell.folate` data for three ventilation conditions

The t statistic was introduced by William Sealy Gosset for cheaply monitoring the quality of beer brews. "Student" was his pen name. Gosset was a statistician for the Guinness brewery in Dublin, Ireland, and was hired due to Claude Guinness's innovative policy of recruiting the best graduates from Oxford and Cambridge to apply biochemistry and statistics to Guinness' industrial processes. Gosset published the t test in Biometrika in 1908, but was forced to use a pen name by his employer who regarded the fact that they were using statistics as a trade secret. In fact, Gosset's identity was unknown not only to fellow statisticians but to his employer—the company insisted on the pseudonym so that it could turn a blind eye to the breach of its rules.

A t test is used in the following sort of situation: You have two sets of data (e.g., responses to a drug trial), each with its mean, standard deviation, and number of measurements. You want to test whether the means are significantly different, or, put another way, to test the *null hypothesis* that the means are equal. The samples may be independent of each other, e.g., individuals randomly assigned to an experimental group or a control group; or they may be paired (dependent), e.g., identical twins, individuals matched on age, or the same individuals examined before and after a drug trial. An equivalent to the two-sample t test is a one-sample test of whether the mean has the value specified in a null hypothesis.

Two qualifications apply to the valid use of a t test. Most importantly, the underlying distributions of the populations from which the samples are drawn can be assumed to be normal. If this assumption is not valid, then a *distribution-free* or *non-parametric* test such as the Wilcoxon (also called Mann-Whitney) should be used. Less importantly, the t test strictly defined assumes that the variances of the populations are equal. If this assumption is not made, a variant called Welch's t test is used; this is the default in R.

The t statistic for a single sample is defined as

$$t = \frac{\bar{x} - \mu}{SEM}; \quad SEM = \frac{\sigma}{\sqrt{n}} \tag{13.1}$$

where SEM is the standard error of the mean if there are n measurements and the variance is σ^2.

For a two-sample t test, the equation for the t statistic is

$$t = \frac{\bar{x_2} - \bar{x_1}}{\sqrt{SEM_1^2 + SEM_2^2}} \tag{13.2}$$

In either case, the calculated t value allows calculation of a p-value for the probability of the null hypothesis from the t distribution (p as a function of t for the appropriate number of degrees of freedom). If p is below the threshold chosen for statistical significance (often the 0.05 level), then the null hypothesis is rejected in favor of an alternative hypothesis that the samples differ or the single sample has a mean different from the hypothesized value. This entire process is automated in R, as we shall see from the examples below.

13.2.1 One-sample t test

We begin with an example of a one-sample t test using simulated data on the effect of glue-sniffing on norepinephrine (NE) concentration in the brains of lab rats. "Measurements" on six rats exposed to the toluene in glue gave

```
> NE = c(427, 637, 517, 575, 493, 541)
```

ng NE per g of rat brain. We first take an overview of the data, and calculate the standard deviation and SEM.

```
> summary(NE)
```

```
      Min. 1st Qu.   Median     Mean 3rd Qu.     Max.
     427.0   499.0    529.0    531.7   566.5    637.0
> sd(NE)
[1] 71.73
> sem = sd(NE)/6; sem
[1] 11.95
```

We then use t.test to ask whether the observed mean is different from the mean of 444 ng/g NE measured in a set of control rats.

```
> t.test(NE, mu=444)


One Sample t-test

data:   NE
t = 2.994, df = 5, p-value = 0.03032
alternative hypothesis: true mean is not equal to 444
95 percent confidence interval:
 456.4 606.9
sample estimates:
mean of x
    531.7
```

This output is to be interpreted as follows.

- t = 2.994 is more than 2 SEMs below the postulated mean μ, indicating a small probability that the sample has such a mean.
- The number of degrees of freedom, df, is 5 although there are measurements on 6 mice. Given 5 measurements and the mean, the 6th can be calculated so is not independent.
- The probability that the mean = μ is small, 0.03032, well below the standard 0.05 criterion.
- The alternative hypothesis is that the true mean is not equal to 444.
- There is a 95% likelihood that the true mean lies between 456.4 and 606.9.
- The sample mean = 531.7, an estimate of the true mean for the population.

The foregoing is a two-tailed t test, asking whether the observed mean is different (either greater or less) than the postulated mean. If your alternative hypothesis was that the amount of norepinephrine in glue-sniffing mice is greater than 444, you would use a one-tailed test with alternative="greater" which can be abbreviated alt="g":

```
> t.test(NE, mu=444, alt="g")


One Sample t-test

data:   NE
t = 2.994, df = 5, p-value = 0.01516
```

```
alternative hypothesis: true mean is greater than 444
95 percent confidence interval:
 472.7    Inf
sample estimates:
mean of x
     531.7
```

You can see other options for t.test by consulting the R Help system: ?t.test.

13.2.2 Two-sample t test

We follow Dalgaard [15, pp. 87–88] in using the energy dataset in his ISwR package. This dataset contains data on the daily energy expenditure (kJ) of 22 women, classified by "stature" as "lean" or "obese":

```
> library(ISwR)
Loading required package: survival
Loading required package: splines

Attaching package: 'ISwR'

> data(energy)
> attach(energy)
> energy
   expend stature
1    9.21   obese
2    7.53    lean
3    7.48    lean
4    8.08    lean
5    8.09    lean
6   10.15    lean
7    8.40    lean
8   10.88    lean
9    6.13    lean
10   7.90    lean
11  11.51   obese
12  12.79   obese
13   7.05    lean
14  11.85   obese
15   9.97   obese
16   7.48    lean
17   8.79   obese
18   9.69   obese
```

```
19    9.68    obese
20    7.58     lean
21    9.19    obese
22    8.11     lean
```

Running t.test, where the tilde indicates that expend is classified by stature, we get

```
> t.test(expend~stature)

Welch Two Sample t-test

data:   expend by stature
t = -3.8555, df = 15.919, p-value = 0.001411
alternative hypothesis: true difference in means is not
  equal to 0
95 percent confidence interval:
 -3.459167 -1.004081
sample estimates:
 mean in group lean mean in group obese
          8.066154            10.297778
```

The default in R is Welch's t test, which does not assume equal variances, which in turn leads to the fractional df. The confidence interval is for the *difference* in the means of the two groups. We see that there is only 0.1411% probability that the means are drawn from the same distribution.

To use a t test in which the variances are assumed equal, we use the option var.equal=T:

```
> t.test(expend~stature,var.equal=T)

Two Sample t-test

data:   expend by stature
t = -3.9456, df = 20, p-value = 0.000799
alternative hypothesis: true difference in means is
  not equal to 0
95 percent confidence interval:
 -3.411451 -1.051796
sample estimates:
 mean in group lean mean in group obese
          8.066154            10.297778
```

The results are not much different, although df now is an integer $(13+9-2=20)$.

13.2.3 Two-sample Wilcoxon test

If the assumption is doubtful that the distributions are normal, you can run a Wilcoxon (sometimes called a Mann-Whitney) test instead of a t test. The Wilcoxon test replaces the data by their rank (without regard to grouping) and calculates the sum of the ranks in one group. The p-value thus calculated is equivalent to sampling (with probability 1/2 at each turn) n_1 values without replacement from the numbers 1 to $n_1 + n_2$.

```
> wilcox.test(expend~stature)

Wilcoxon rank sum test with continuity correction

data:   expend by stature
W = 12, p-value = 0.002122
alternative hypothesis: true location shift is not
   equal to 0

Warning message:
cannot compute exact p-value with ties in:
   wilcox.test.default(x = c(...
```

W is the sum of ranks in the first group minus its theoretical minimum (0). Since two of the values are tied, the p-value calculated from the ranks must be estimated using a normal approximation.

The Wilcoxon test is particularly useful if the data are rank ordered in the first place.

13.2.4 Paired t test

A paired t test uses two samples matched in some way. An example [52, p. 356] is from an experiment [11] that studied the effects of progesterone on the cAMP content of maturing *Xenopus laevis* oocytes. Oocytes from each of four female frogs were divided into two batches, one exposed to progesterone and one not. The eggs were assayed for cAMP (pmol/oocyte) after 2 minutes with the following results:

```
> control = c(6.01, 2.28, 1.51, 2.12)
> progesterone = c(5.23, 1.21, 1.40, 1.38)
> mean(control)
[1] 2.98
> mean(progesterone)
[1] 2.305

> t.test(control, progesterone, paired=T)
```

```
Paired t-test

data:   control and progesterone
t = 3.339, df = 3, p-value = 0.04443
alternative hypothesis: true difference in means is
   not equal to 0
95 percent confidence interval:
 0.03159 1.31841
sample estimates:
mean of the differences
              0.675
```

It appears that the null hypothesis is just barely ruled out at the 0.05 level. If a less stringent cutoff (e.g., 0.1) were applied, the alternative hypothesis would be more readily accepted. Note that the 95% confidence interval is for the difference of the means.

13.2.5 Statistical power calculations

There are two types of errors in statistical testing of hypotheses:

- Type I: A correct hypothesis is rejected
- Type II: An incorrect hypothesis is accepted

The power of a statistical test is the probability that the test will reject a false null hypothesis (that it will not make a Type II error). As power increases, the chance of a Type II error decreases, and vice versa. The probability of a Type II error is referred to as β. Therefore power is equal to $1 - \beta$. Typically, statistical studies aim for powers of 0.8 or 0.9. (See the article in *Wikipedia*.) Often the goal is to calculate the sample size n needed to achieve a given power if it assumed that the "true difference" delta and standard deviation sd are known, and a significance level (sig.level or p-value) is required.

As an example, suppose you wanted to investigate the effect of a food additive on the growth rate of baby mice, measured by weight of experimental vs. control at 2 weeks after birth. You want to know how many mice must be in each group, given that you can detect a difference in weight of 1 g and that the weight in the population has a standard deviation of 3 g, if you are to have a power of 90% in a two-sided test at the 5% level.

```
> power.t.test(delta=1,sd=3,sig.level=0.05,power=0.9)

        Two-sample t test power calculation

            n = 190.1
```

```
        delta = 1
           sd = 3
    sig.level = 0.05
        power = 0.9
  alternative = two.sided
```

```
 NOTE: n is number in *each* group
```

If you had only 150 in each group, the power would be

```
> power.t.test(delta=1, sd=3, sig.level=0.05, n=150)
```

```
    Two-sample t test power calculation
```

```
            n = 150
        delta = 1
           sd = 3
    sig.level = 0.05
        power = 0.8206
  alternative = two.sided
```

```
 NOTE: n is number in *each* group
```

power.t.test takes any four of the five variables, and calculates the fifth.
The default for power.t.test is a two-sided t test. If the hypothesis to be tested
is that the additive increases growth rate, then one would use a one-sided test:

```
> power.t.test(delta=1,sd=3,sig.level=0.05,
  power=0.9,alt="one.sided")
```

```
    Two-sample t test power calculation
```

```
            n = 154.8
        delta = 1
           sd = 3
    sig.level = 0.05
        power = 0.9
  alternative = one.sided
```

```
 NOTE: n is number in *each* group
```

so only 155 mice in each group would be needed, instead of 190 in a two-sided test.
   Two other variations in the power calculation are type="one.sample" and
type="paired". For example, suppose we had paired subjects (e.g., twins from
the same litter) in the food additive test:

```
> power.t.test(delta=1, sd=3, sig.level=0.05,
```

```
power=0.9, type="paired")

    Paired t test power calculation

              n = 96.5
          delta = 1
             sd = 3
      sig.level = 0.05
          power = 0.9
    alternative = two.sided

NOTE: n is number of *pairs*, sd is std.dev. of
   *differences* within pairs
```

13.3 Analysis of spectral data

We have covered some of the basics of data fitting in Chapter 5 on "Equilibrium and Steady State Calculations", but there are a number of additional aspects that we will examine in this section. In particular, a common problem is to disentangle overlapping individual contributions to spectra or exponential decays in the presence of noise and (often) tilted and/or curved baselines. Our basic tool is the R function nls, to "[d]etermine the nonlinear (weighted) least-squares estimates of the parameters of a nonlinear model."

13.3.1 Fitting to a sum of decaying exponentials

We begin with a problem that occurs in many fields of science: fitting an observed multi-component decay to a sum of exponentials. In biophysics and molecular biology, this arises most commonly in fluorescence decay experiments.

To have some simulated data to work with, we generate a function that's the sum of two decaying exponentials plus random noise. The time scale might be ns in a fluorescence decay experiment, but that is irrelevant to our example. Each exponential contribution to the decay is characterized by an amplitude A and a decay time τ.

```
> t = seq(0,20,.5)
> y = 3*exp(-t/3)+exp(-t/10)+0.05*rnorm(length(t))
```

We use nls to fit to a two-exponential model.

```
> yfit = nls(y ~ A1*exp(-t/tau1) + A2*exp(-t/tau2),
    start=c(A1=2, A2=2,tau1=4, tau2=10))
> yfit
```

```
Nonlinear regression model
  model:   y ~ A1 * exp(-t/tau1) + A2 * exp(-t/tau2)
   data:   parent.frame()
   A1    A2 tau1 tau2
2.92 1.15 2.86 9.23
  residual sum-of-squares: 0.11

Number of iterations to convergence: 12
Achieved convergence tolerance: 8.4e-06
```

The parameters are available as a vector in `coef(yfit)`. We extract the coefficents and calculate the fitting curve and residuals.

```
> A1 = coef(yfit)[1]
> A2 = coef(yfit)[2]
> tau1 = coef(yfit)[3]
> tau2 = coef(yfit)[4]
> y.fit = A1*exp(-t/tau1) + A2*exp(-t/tau2)
> yres = y-y.fit
```

We then plot the points, the fitting curve, and the residuals.

```
> par(mfrow=c(1,2))
> plot(t,y)
> lines(t,y.fit)
> plot(t,yres)
> abline(h=0)
> par(mfrow=c(1,1))
```

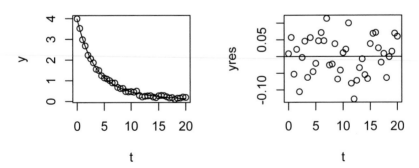

Fig. 13.10 Nonlinear least-squares fit (left) and residuals (right) to a function of two decaying exponentials plus random noise

The fit looks reasonably good and the residuals appear randomly distributed about 0.

The `summary` function in `nls` yields more detail on the estimates and standard errors of the parameters, as well as the probabilities that the estimates are significant.

```
> summary(yfit)

Formula: y ~ A1 * exp(-t/tau1) + A2 * exp(-t/tau2)

Parameters:
       Estimate Std. Error t value Pr(>|t|)
A1       2.919      0.392    7.44   7.4e-09 ***
A2       1.148      0.407    2.82    0.0077 **
tau1     2.863      0.306    9.37   2.6e-11 ***
tau2     9.225      1.902    4.85   2.2e-05 ***
---
Signif. codes:  0 *** 0.001 ** 0.01 * 0.05 . 0.1   1

Residual standard error: 0.0545 on 37 degrees of
freedom

Number of iterations to convergence: 12
Achieved convergence tolerance: 8.4e-06
```

13.3.2 Fitting a superposition of Gaussian spectra

In the UV and visible spectroscopy of proteins, nucleic acids, and other biomolecules, we often encounter overlapping spectral bands. Such spectra may have a significant amount of noise, depending on molecular concentrations and the characteristics of the spectrophotometer. As the far UV is approached, the baseline may also increase. We show how to use `nls` to determine the individual spectral components in such a complex spectrum.

We begin by assuming a flat baseline. We first define the Gaussian shape that is typical of individual electronic spectra, generate the superposition spectrum of two individual bands with representative noise, and plot it.

```
> gauss = function(A,x0,sd,x) {A*exp(-(x-x0)^2/(2*sd^2))}
> x = 200:280 # A typical UV spectral range, in nm
> y = gauss(1,237,8,x) + gauss(2,222,5,x) + rnorm(length(x),
  0,0.05)
> plot(x,y, type="l")
```

From inspection of the spectrum, we make some starting guesses about the parameters and run `nls`. indexnls

```
> yfit = nls(y~gauss(A1,x1,sd1,x)+gauss(A2,x2,sd2,x),
    start=c(A1=0.8,x1=240,sd1=5,A2=1.5,x2=225,sd2=5))
```

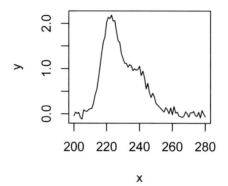

Fig. 13.11 A typical UV spectrum consisting of two overlapping Gaussian peaks plus random noise

```
summary(yfit)

Formula: y ~ gauss(A1, x1, sd1, x) + gauss(A2, x2,
sd2, x)

Parameters:
      Estimate Std. Error t value Pr(>|t|)
A1     0.99167    0.01919   51.67   <2e-16 ***
x1   237.21105    0.46240  513.00   <2e-16 ***
sd1    7.92454    0.37008   21.41   <2e-16 ***
A2     2.04141    0.04635   44.05   <2e-16 ***
x2   222.09743    0.12992 1709.44   <2e-16 ***
sd2    4.97295    0.10202   48.74   <2e-16 ***
---
Signif. codes:  0 *** 0.001 ** 0.01 * 0.05 . 0.1   1

Residual standard error: 0.05083 on 75 degrees of
freedom

Number of iterations to convergence: 6
Achieved convergence tolerance: 8.503e-07
```

The fitting parameters are close to the actual values. We extract the fitted parameters, calculate the fitted spectrum, and superimpose it on the experimental points. Then we calculate and plot the residuals.

```
> A1 = coef(yfit)[1]; x1 = coef(yfit)[2]; sd1 = coef(yfit)[3]
> A2 = coef(yfit)[4]; x2 = coef(yfit)[5]; sd2 = coef(yfit)[6]
> y.fit = gauss(A1,x1,sd1,x) + gauss(A2,x2,sd2,x)
```

```
> par(mfrow = c(1,2))
> plot(x,y)
> lines(x,y.fit)
> yres = y - y.fit
> plot(x,yres); abline(h=0)
> par(mfrow = c(1,1))
```

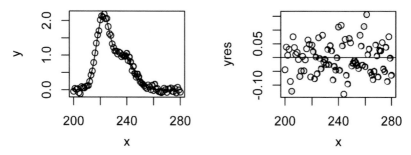

Fig. 13.12 Resolution of the spectrum into its components by nonlinear least-squares fitting (left) with residuals (right)

Now let us add a quadratic baseline that tilts upward as we go to shorter wavelengths, starting at 0 at 280 nm and approaching 1 at 200 nm. The equation for the baseline is $bl = 0.2 * (x - 280) + 1.5e - 4 * (x - 280)^2$. Repeating our previous procedure,

```
> bl = 0.0025*(x-280) + 1.5e-4*(x-280)^2
> y = gauss(1,237,8,x) + gauss(2,222,5,x) +
      rnorm(length(x),0,0.05) + bl

> yfit = nls(y~gauss(A1,x1,sd1,x)+gauss(A2,x2,sd2,x) +
      b+m1*x+m2*x^2 ,
      start=c(A1=0.8,x1=240,sd1=5,A2=1.5,x2=225,sd2=5,
      b=0, m1=.1, m2=0.001))
> summary(yfit)

Formula: y ~ gauss(A1, x1, sd1, x) + gauss(A2, x2,
    sd2, x) + b + m1 * x + m2 * x^2

Parameters:
       Estimate Std. Error  t value Pr(>|t|)
A1    1.017e+00  3.253e-02   31.281  < 2e-16 ***
x1    2.364e+02  5.317e-01  444.607  < 2e-16 ***
sd1   8.150e+00  5.027e-01   16.215  < 2e-16 ***
```

```
A2     1.991e+00   5.953e-02    33.449   < 2e-16 ***
x2     2.219e+02   1.420e-01  1562.622   < 2e-16 ***
sd2    4.881e+00   1.138e-01    42.891   < 2e-16 ***
b     -4.318e+01   1.384e+00   -31.199   < 2e-16 ***
m1     1.070e-01   1.183e-02     9.044  1.73e-13 ***
m2     1.692e-04   2.482e-05     6.815  2.42e-09 ***
---
Signif. codes:  0 *** 0.001 ** 0.01 * 0.05 . 0.1     1

Residual standard error: 0.0502 on 72 degrees of
freedom

Number of iterations to convergence: 7
Achieved convergence tolerance: 1.554e-06
```

We extract the coefficients as before, and get fitting curves with both baseline included and baseline subtracted.

```
> A1 = coef(yfit)[1]; x1 = coef(yfit)[2]; sd1 = coef(yfit)[3]
> A2 = coef(yfit)[4]; x2 = coef(yfit)[5]; sd2 = coef(yfit)[6]
> b = coef(yfit)[7]; m1 = coef(yfit)[8]; m2 = coef(yfit)[9]
> y.fit = gauss(A1,x1,sd1,x) + gauss(A2,x2,sd2,x) + b + m1*x +
  m2*x^2
> y.baselinecorrected = gauss(A1,x1,sd1,x) + gauss(A2,x2,sd2,x)
```

We plot the full fit with residuals, and then the baseline-corrected spectrum. We find that `nls` does an excellent job of cleaning up the spectrum.

```
> par(mfrow = c(2,2))
> plot(x,y)
> lines(x,y.fit)
> yres = y - y.fit
> plot(x,yres); abline(h=0)
> plot(x,y.baselinecorrected, type="l")
> par(mfrow = c(1,1))
```

13.3.3 *Processing mass spectrometry data*

Baseline subtraction and peak identification are common problems in all sorts of spectroscopy, and mass spectrometry of proteins—the technique that underlies proteomics—is no exception. In fact, given the large number of peaks, noisy signals, and often highly nonlinear baseline in mass spectrometry (MS), these problems may be more severe than in most other forms of spectroscopy. Therefore, specialized software packages have been written to treat MS data.

In this section we consider one such software package, PROcess, written in R as part of the Bioconductor project. The package, written and maintained by Xiaochun

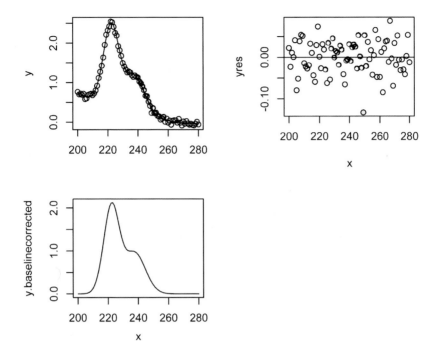

Fig. 13.13 (top left) Resolution of the two-Gaussian spectrum on a quadratic baseline; (top right) Residuals after `nls` fitting; (bottom) spectrum with baseline subtracted

Li, is available at

http://bioconductor.org/packages/2.2/bioc/html/PROcess.html.

It contains both processing software and sample data sets to which the software may be applied. `PROcess` deals with a specific type of mass spectrometry, SELDI-TOF-MS. which stands for "surface enhanced laser desorption/ionization time-of-flight mass spectrometry". However, the spectrum-processing principles are common to most forms of MS.

You can install `PROcess` by typing

```
> source("http://bioconductor.org/biocLite.R")
> biocLite("PROcess")
```

at your R console while connected to the Internet. An introduction to how to use `PROcess` is available as a vignette on the web page, and also as a chapter by Li et al. in the book edited by Gentleman et al. [27, Ch. 6]. Those sources should be consulted for details and for a more complete survey of the spectral analysis process. We follow the sections on baseline subtraction and peak detection from that chapter and vignette closely, and use the code samples verbatim.

To get an idea of the nature of the data analysis problem, load PROcess and access one of the sample spectra by the following R code:

```
> library(PROcess)
> fdat = system.file("Test", package="PROcess")
> fs = list.files(fdat, pattern = "\\.*csv\\.*",
  full.names=TRUE)
> f1 = read.files(fs[1])
> plot(f1,type="l",xlab="m/z")
> title(basename(fs[1]))
```

R will tell you that it is loading the required packages Icens, survival, and splines, and then plot spectral intensity vs. m/z (mass/charge) for sample f1.

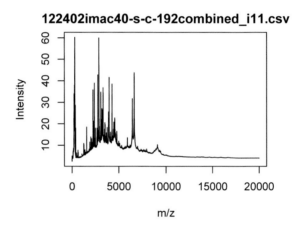

Fig. 13.14 Typical mass spectrum of a protein

The elevated, inconstant baseline, due mainly to chemical noise from the energy-absorbing molecules in the surface and to overloading the detector with ions, is immediately apparent. The baseline should be leveled to zero if different mass spectra are to be compared. This is done in PROcess by subtracting from the observed spectrum an estimate of its "bottom", obtained by fitting a local regression line to the local minima. The following code calls that procedure, and plots the raw and process spectra and the calculated baseline:

```
> bseoff = bslnoff(f1,method = "loess", bw=0.1,
  xlab = "m/z", plot=T)
> title(basename(fs[1]))
```

As Li et al. [27, Ch. 6] note, "Caution needs to be employed in choosing the bandwidth [bw]; too small a bandwidth may result in the removal of some peaks, especially wide short peaks, which may be potentially important."

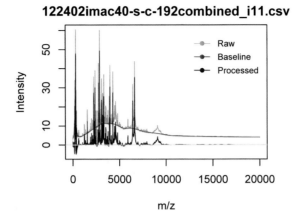

Fig. 13.15 Mass spectrum with baseline removal

Once the baseline is leveled and subtracted, the next task is to identify the significant peaks. This is done by smoothing over a moving window of carefully chosen width, and then locating the local maxima of the smoothed spectra. The code that carries this out is called by

```
> pkgobj = isPeak(bseoff, span=81, sm.span=11,
> plot=T,zerothrsh=2,area.w=0.003,ratio=0.2,main="a)")
```

with the result shown in Figure 13.16.

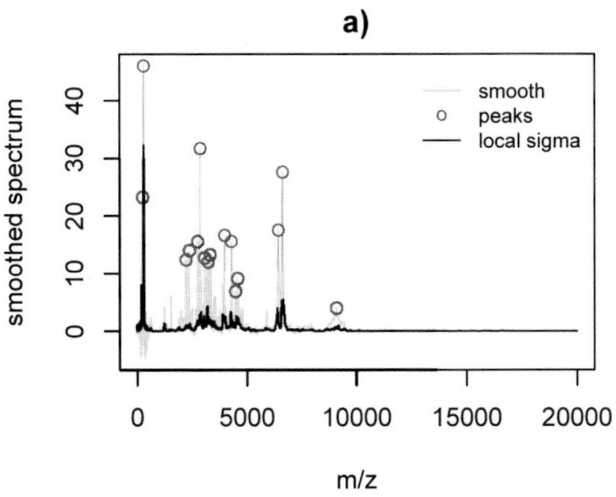

a)

Fig. 13.16 Identification of mass spectral peaks

Baseline subtraction and peak identification are the necessary prerequisites to comparing a given spectrum to a set of others, with the goal of identifying those proteins or peptides (which correspond to peaks) that are differentially expressed under various conditions such as disease states. The comparison involves further steps such as m/z alignment, peak amplitude standardization, cutoff of overly noisy regions especially at low m/z, and quality assessment of the spectrum. (Is it too noisy? Does it have enough peaks to be interesting?), and peak quantitation. Alignment of peaks across spectra is carried out with "proto-biomarkers", peaks that are likely to represent the same protein in different spectra. We will not deal with these subsequent steps, but refer the reader to the PROcess vignette or to the chapter by Li et al. [27, Ch. 6].

13.4 Problems

1. Here are simulated data on the results (weight.gain) of treating three groups of mice (genotypes *AA*, *Aa*, and *aa*) with four levels of an experimental diet ("zero","low", "medium", "high").

   ```
   > weight.gain = matrix(c(65,150,60,24,22,33,11,7,4,
     5,4,2), nrow=3, byrow=T)
   > weight.gain
        [,1] [,2] [,3] [,4]
   [1,]   65  150   60   24
   [2,]   22   33   11    7
   [3,]    4    5    4    2
   ```

   The rows denote the genotype, the columns the diet level.
   Add row and column names to weight.gain to give a properly labeled table, and show the table.
   The transpose of a matrix M is denoted in R by t(M). Show the table obtained from t(weight.gain).

2. We might want to visualize the t(weight.gain) data as a barplot, with either the genotype or diet level as the independent variable. By default, barplot yields a stacked plot, while the option beside=TRUE (which can be abbreviated to beside=T) gives the bars side by side. Using the par(mfrow=c(2,2)) instruction to plot a 2 × 2 array of graphs, plot stacked and unstacked bar plots of the weight.gain data with diet level as the independent variable on the top row and genotype on the bottom row.

3. The margin.table function is an easy way to compute the sum of entries for a given factor in a table. Look up margin.table in R Help, and apply it to the rows and columns of weight.gain. Then plot side-by-side barplots in which diet level and genotype are the independent variables.

4. Other kinds of plots are sometimes useful. Look up dotplot in R Help and apply it to t(weight.gain). Which barplot in Problem 2 is the result most similar to?

Pie charts are not generally useful in scientific work, but with the `weight.gain` data they can be informative. Try out the following code adapted from [15, p. 79]:

```
> opar=par(mfrow=c(2,2),mex=0.8,mar=c(1,1,2,1))
> slices=c("white","grey80","grey50","black")
> pie(weight.gain["AA",],main="AA",col=slices)
> pie(weight.gain["Aa",],main="Aa",col=slices)
> pie(weight.gain["aa",],main="aa",col=slices)
> par(opar)
```

Dalgaard [15, p. 71] explains the "par magic" as follows: "The mex setting reduces the interline distance and mar reduces the number of lines that surround the plot region. All of the original values of the changed settings can be stored in a variable (here `opar`) and reestablished with `par(opar)`." Check this out by omitting the `opar/par(opar)` lines and just using `par(mfrow=c(2,2)`.

5. Plot a histogram for the \texttt{react} react data set in ISwR. Compare it with the result of `truehist` from the `MASS` package. Read the descriptions of `hist` and `truehist` to see why the results are different.

6. Use a t test with the data set `vitcap` in ISwR to compare the vital capacities of the two groups. What is the 99% confidence interval for the difference? Why might the result of this comparison be misleading? Next do the analysis using the nonparametric Wilcoxon test. Do the t and Wilcoxon tests give notably different results? (Adapted from [15, Ch. 4 Exercises].)

7. R contains several simple but useful graphical functions for inspection of data. Among these are `stem` and `stripchart`. Look up these functions in R Help to see how they work. Generate a vector of 100 normally-distributed random numbers and apply `stem` and `stripchart`. Compare the default and "jitter" methods of `stripchart`.

8. Generate simulated data for a two-exponential decay in which the decay times are separated by a factor of 2 (e.g., 5 and 10 ns) and the decays are of equal amplitude, in the presence of 5% normally distributed random noise. Do you get good results? Now try it with three exponential decays (e.g., 5, 10, and 20 ns).

9. Add a third Gaussian spectral component to the two in the example with the quadratic baseline and 5% noise, with amplitude 0.5, $x0 = 230$, and $sd = 4$, and run the code to extract the parameters of the three spectral components. In an additional plot in the lower right of the 2×2 array of plots, plot the original spectral components in black and the fitted components in red.

10. Redo the `PROcess` mass spectrometry calculations with component 2 of `fs`.

Chapter 14
Microarrays

14.1 Introduction

DNA microarrays are one of the key new technologies in biology. A microarray is a solid surface, typically a microscope slide or silicon chip, to which short DNA molecules (*probes*) are bonded in a regular array of thousands of tiny spots. Each spot, or feature, has a specific DNA sequence and is located at a known place in the array. The probes are designed to bond by Watson-Crick hybridization to the DNA molecules in a *target* sample which typically consists of either cDNA reverse transcribed from mRNA, or genomic DNA. Hybridization of the target to the probes is usually detected and quantified by fluorescence from the fluorescently labeled target molecules. The result is a quantitative measure of the relative abundance of DNA sequences in the target. This information can be used in a variety of ways, including

- measurement of changes in gene expression levels, e.g., in response to illness or developmental stage, by measuring the relative abundance of cDNAs from mRNAs;
- detection of single nucleotide polymorphisms (SNPs) that may be indicators of susceptibility to disease or useful in forensic analysis;
- comparison of genome content of different cells or closely related organisms;
- detection of alternative splicing in DNA transcription.

Three general types of goals for microarray studies have been distinguished [56, p. 3 and Chs. 7–9]. In *class comparison* studies, predefined classes exist (e.g., different tissue types or disease states) and the goal is to determine which genes are differentially expressed in each class. In *class prediction* studies, the goal is to find gene expression patterns among different predefined classes (e.g., patients who do or don't respond to a particular drug treatment) that enable prediction of the class to which a new member of the population will belong. In *class discovery* studies, the goal is to discover new taxonomies, or groups of co-expressed and co-regulated genes, based on expression profiles. Each of these types of studies requires somewhat different experimental design and statistical analysis.

V. Bloomfield, *Computer Simulation and Data Analysis in Molecular Biology and Biophysics,*   279
Biological and Medical Physics, Biomedical Engineering, DOI: 10.1007/978-1-4419-0083-8_14,
© Springer Science + Business Media, LLC 2009

A microarray may contain ten thousand or more spots, and therefore can carry out thousands of comparative genetic analyses at once. (Spots are often replicated on an array to provide internal calibration.) This enormous amount of information can provide great insight into genetic regulatory processes, but it also poses great challenges to data quality and adequate statistical analysis. These issues are the subject of this chapter, but we can provide only a brief overview. More detailed treatments can be found in the books by Hahne et al. [31], Gentleman et al. [27], Stekel [60], and Simon et al. [56], as well as in many reviews and manufacturers' literature.

14.2 Preprocessing overview

There are two main types of microarrays: two-color spot arrays and high-density oligonucleotide arrays. The two-color spot platforms, which were the original microarrays, use target DNA labeled with either red or green fluorophores to give two sets of data on a single array. Background noise is determined from areas of the glass slide that do not contain probe. The high-density oligonucleotide platform, of which the Affymetrix GeneChip^{TM} is the most common, has two types of probes on a single chip: a *perfect match* 25-mer intended to hybridize under stringent conditions with a given gene sequence, and a *mismatch* 25-mer with the complementary base at position 13 intended to measure nonspecific hybridization.

Regardless of the type of array platform, there are six preprocessing steps that must take place before the probe-level spot intensities can be converted into measures of expression for further analysis [27, Ch. 1].

Image analysis: A scanner measures the fluorescence intensity signal at each point on the array. The scan distance may be about 3 μm to produce one pixel; but the typical probe dimension is about 50 μm, so each probe feature is made up of many pixels. The pixel intensities must be aggregated and analyzed to give probe-level data.

Data import: Image data and other descriptions of the experiment (samples, genes, probes, etc.) need to be assembled and and organized.

Background adjustment: Some of the observed intensity is due to nonspecific binding of target to the array. This background must be subtracted to give a suitable measure of specific binding.

Normalization: Microarray studies involve comparisons between arrays or differently labeled samples on the same array. There are many potential sources of variation that must be taken into account before the comparisons can be useful for the study of differential gene expression. These sources of variation can include differences in efficiencies of reverse transcription of mRNA into cDNA, labeling of the target DNA, and hybridization of target to probe; local properties of the array surfaces; freshness of the reagents; and differences between operators. Normalization attempts to take these sources of variation into account.

Summarization: Most arrays have several probes directed against the same target. Even after background adjustment and summarization, the probes are likely to give several different readings of the amount of target. The readings must therefore be combined to give a single expression value.

Quality assessment: Some of the measurements may exhibit fluctuations well beyond those expected from random fluctuations. The data must be screened so that these divergent measurements can be flagged and either dropped from the study or given low weights.

It is important to note that the validity of the background adjustment and normalization steps rest on a fundamental assumption: that only a few of the many transcripts in the target solution are significantly up- or down-regulated. These few, which are ultimately the transcripts of interest, are too small a proportion of the whole to affect the average behavior of the array. Thus, all arrays that are prepared the same way should have the same statistical properties. The background adjustment and normalization steps are intended to bring the arrays to this properly matched state, so that the significantly different targets can be confidently identified.

In the following sections we show how these preprocessing steps can be applied to high-density oligonucleotide GeneChip^{TM} arrays. Preprocessing of two-color arrays involves similar procedures. Affymetrix offers a comprehensive suite of software to accomplish the data acquisition and analysis steps, but it is a closed system. We shall use some of the tools in the `affy` package developed by the Bioconductor project, which display the individual steps in a pedagogically useful way and provide statistically important alternatives for most of the steps. The Bioconductor software does not include image analysis, however, and we shall use examples in which the data have already been imported, so our treatment begins with the background adjustment step. Our examples come from Chapters 1–3 of the book by Gentleman et al. [27].

14.3 The `affy` package

To begin, go to the Bioconductor web site `www.bioconductor.org` and choose `Packages -> Software -> affy`. There are two vignettes you should download and have available for reference: `affy.pdf` and `builtinMethods.pdf`. Alternatively, you can read them from the `openVignettes()` command in `affy` (see below). To install the `affy` package itself on your computer, start R and enter:

```
source("http://bioconductor.org/biocLite.R")
biocLite("affy")
```

Then open `affy` with `> library(affy)`. You will get the following introductory material:

```
Loading required package: Biobase
```

```
Loading required package: tools

Welcome to Bioconductor

  Vignettes contain introductory material. To view, type
  'openVignette()'. To cite Bioconductor, see
  'citation("Biobase")' and for packages 'citation(pkgname)'.

Loading required package: affyio
Loading required package: preprocessCore
```

This might be a good time to use the `openVignette()` command to read item 7 (Bioconductor Overview) and skim item 1 (Primer: Description of affy). You will want to refer back to the latter as we proceed through the chapter.

The fundamental entity in `affy` is the `AffyBatch class`. In the "Description of affy" vignette, Sec. 6.1, we read "The AyBatch class has slots to keep all the probe level information for a batch of Cel files, which usually represent an experiment. It also stores phenotypic and MIAME information." A CEL file is the result of processing the raw image data files to give estimated probe intensities. MIAME stands for "Minimum Information About a Microarray Experiment".

To get an idea of the sort of information contained in an `AffyBatch` object, load the `ALLMLL` package and access the data in `MLL.B`

```
> library(ALLMLL)
> data(MLL.B)
```

Then `?MLL.B` gives

AffyBatch instances MLL.A and MLL.B

Description

These AffyBatch objects contain a subset of arrays from a large acute lymphoblastic leukemia (ALL) study,

Usage

data(MLL.A) data(MLL.B)

Format

Each are AffyBatch containing 20 arrays.

Source

This package provides probe-level data for 20 HGU133A and 20 HGU133B arrays which are a subset of arrays from a large ALL study. The data is for the MLL arrays. This data was published in:

Mary E. Ross, Xiaodong Zhou, Guangchun Song, Sheila A. Shurtleff, Kevin Girtman, W. Kent Williams, Hsi-Che Liu, Rami Mahfouz, Susana C. Raimondi, Noel Lenny, Anami Patel, and James R. Downing (2003) Classification of pediatric acute lymphoblastic leukemia by gene expression profiling Blood 102: 2951-2959

The `pData` function "returns a data frame with samples as rows, variables as columns."

```
> pData(MLL.B)
```
                            sample

```
JD-ALD009-v5-U133B.CEL          1
JD-ALD051-v5-U133B.CEL          2
      . . .
JD-ALD520-v5-U133B.CEL         20
```

Names of the genes and probes can be obtained with `geneNames` and `probeNames`. To find out how many there are, use the `length` function:

```
> length(probeNames(MLL.B))
[1] 249502
> length(geneNames(MLL.B))
[1] 22645
```

There are so many that we wouldn't want to print out complete lists, but we can get a subset, e.g., the first ten, with

```
> geneNames(MLL.B)[1:10]
 [1] "200000_s_at" "200001_at"   "200002_at"
     "200003_s_at" "200004_at"   "200005_at"
 [7] "200006_at"   "200007_at"   "200008_s_at"
     "200009_at"
> probeNames(MLL.B)[1:10]
 [1] "200000_s_at" "200000_s_at" "200000_s_at"
     "200000_s_at" "200000_s_at" "200000_s_at"
 [7] "200000_s_at" "200000_s_at" "200000_s_at"
     "200000_s_at"
```

Note that all of the first 10 probes are `"200000_s_at"`, showing the multiplicity of probes for a given gene in the high-density oligonucleotide array. In fact, there are actually 11 probes for the gene on this chip:

```
> which(geneNames(MLL.B)=="200000_s_at")
[1] 1
> which(probeNames(MLL.B)=="200000_s_at")
 [1]  1  2  3  4  5  6  7  8  9 10 11
```

The intensities of the "perfect match" spots are obtained with the pm function, those of the "mismatch" spots with mm:

```
> pm(MLL.B)[1:10]
 [1]  661.5  838.8  865.3  425.8  986.8  612.3  627.5
      243.3  293.3 1117.5
> mm(MLL.B)[1:10]
 [1] 250.5 218.0 134.3 190.5 236.3 195.3 129.5 120.5
     143.5 173.5
```

In this case the mismatch intensities are all smaller than the perfect match intensities, as should be the case. However, we shall soon see that this is not always the case, so that more complex methods of subtracting background must be employed.

14.4 Importing data not in standard form

The Bioconductor packages are set up to receive and process data in the form of .CEL files and other standard ouput files from the microarray manufacturers. If your data and annotations are not in such a form, you can assemble them into a Bioconductor *ExpressionSet* in the manner described by Hahne et al. [31, pp. 12–17].

The most obvious data needed are the expression values. Typically, they will be in a tabular spreadsheet format, with F rows, one for each feature, and S columns, one for each sample. There will also likely be headers for each column. The data should be exported from the spreadsheet as a tab-separated file, and imported into R using the read.table command with header=TRUE. The imported table can then be turned into a matrix with as.matrix. Usually considerable annotation will also be desired, to provide information about the samples and the features and to include a description of the experiment. Ways to include this additional information in the ExpressionSet are described in [31]. However, a minimal set containing only the expression data can be created with

```
minimalSet = new("ExpressionSet", exprs = expressionData)
```

where expressionData is the name of the matrix formed from the imported spreadsheet table.

14.5 The need for preprocessing

To see an example of the sort of spatial artifact that can occasionally afflict an array, execute the following code once you've loaded affy:

```
> library(ALLMLL)
> data(MLL.B)
> ProblemChip = MLL.B[,2]
> palette.gray = c(rep(gray(0:10/10),
    times=seq(1,41,by=4)))
> image(ProblemChip,col=palette.gray)
```

The intensity profile on the second MLL.B array has been plotted in grayscale using the image function. Uneven labeling is evident, and is made more prominent by the default log scale of the image function when applied to an AffyBatch object. This example makes clear the importance of quality assessment when doing microarray studies.

A typical illustration of the sorts of adjustments that need to be made to a set of arrays before they can be sensibly analyzed comes from this comparison of eight of the twenty MLL.B chips. (When applied to an AffyBatch object like Data, the hist function yields a smooth density curve for each sample in the object, and the boxplot function displays boxplots for each sample.)

JD-ALD051-v5-U133B.CEL

Fig. 14.1 Spatial artifact due to uneven labeling on a microarray chip

```
> Data = MLL.B[, c(2,1,3:5,14,6,13)]
> sampleNames(Data) = letters[1:8]
> par(mfrow=c(1,2))
> hist(Data, lty=1:8, xlab="Log (base 2) intensities")
> legend(12,1, letters[1:8], lty=1:8, bty="n")
> boxplot(Data)
> par(mfrow=c(1,1)
```

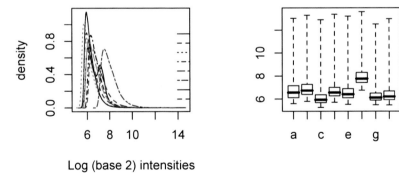

Fig. 14.2 Density curves (left) and boxplots (right) for selected MLL.B chips before preprocessing

In the left-hand figure, a is the ProblemChip we looked at above, as we can see from the bimodal intensity peak. But note that all of the chips start with intensities of approximately $2^6 = 64$, while if the backgrounds were completely clean, they should start at zero. The boxplot function gives further evidence of unequal means and variances, and also emphasizes the anomalously high intensity of sample f.

14.6 Preprocessing steps

14.6.1 The Dilution dataset

To see how the various preprocessing steps work, and what their options are, we'll use the `Dilution` dataset in library `affydata`.

```
> library(affydata)
> data(Dilution)
```

The help page for `Dilution` tells us that it is an "AffyBatch-class containing 4 arrays" whose source is

> Two sources of cRNA A (human liver tissue) and B (Central Nervous System cell line) have been hybridized to human array (HGU95A) in a range of proportions and dilutions. This data set is taken from arrays hybridized to source A at 10.0 and 20 g. We have two replicate arrays for each generated cRNA. Three scanners have been used in this study. Each array replicate was processed in a different scanner.
>
> For more information see Irizarry, R.A., et al. (2001) http://www.biostat.jhsph.edu/ ri-rizarr/papers/index.html

To learn more about `Dilution`, we simply type its name.

```
> Dilution
AffyBatch object
size of arrays=640x640 features (27210 kb)
cdf=HG_U95Av2 (12625 affyids)
number of samples=4
number of genes=12625
annotation=hgu95av2
notes=
```

`hgu95av2` stands for Affymetrix Human Genome U95 Set Annotation Data.

14.6.2 Preliminary inspection

A preliminary `boxplot(Dilution)` inspection gives
    Differences between the centers and dispersions of the distributions show the need for background adjustment and normalization.
    We can look at the expression for a particular probeset, `1001_at` in this case, either by probe number or across the four arrays.

```
> par(mfrow=c(1,2))
> matplot(pm(Dilution,"1001_at"),type="l",col=rep(1,4),
  xlab="Array #", ylab="PM probe intensity")
> matplot(t(pm(Dilution,"1001_at")),type="l",
  col=rep(1,4), xlab="Array #",ylab="PM probe intensity")
> par(mfrow=c(1,1))
```

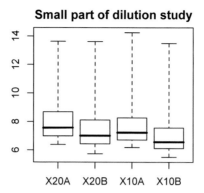

Fig. 14.3 Boxplots of four `Dilution` arrays before background adjustment and normalization

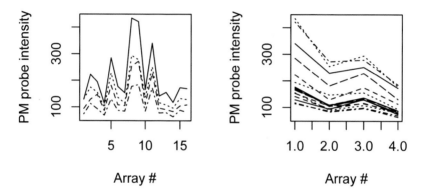

Fig. 14.4 Expression of `Dilution` probeset 1001_at by probe number (left) and across the four arrays (right)

Note that the transpose (`t`) function gives the intensity by array number for the different probes, while the untransposed matrix gives the intensity by probe number for the different arrays.

14.6.3 MAplot

A useful measure of data quality is the `MAplot`. Let Y_{1j} be the log (base 2) expression of gene j and Y_{2j} be the $\log_2$ median expression of that gene across all the arrays in the `AffyBatch`—the "pseudo-median reference chip". (In a two-color array the variables would be red and green intensities.) If every gene were expressed to the same extent for each probe on each array, then a plot of Y_{1j} vs Y_{2j} for all j would be a straight line at 45 degrees. If we let

$$M_j = Y_{1j} - Y_{2j} \tag{14.1}$$

$$A_J = \frac{Y_{1j} + Y_{2j}}{2} \tag{14.2}$$

then M is the difference in $\log_2$ expression values (which equals the log of the twofold expression ratio) and A is the average $\log_2$ expression value. A plot of M vs. A should be parallel to the A axis and the points should lie scattered normally around the line $M = 0$. If we make such an MAplot with the untreated Dilution data we get

```
> MAplot(Dilution, cex=0.75
```

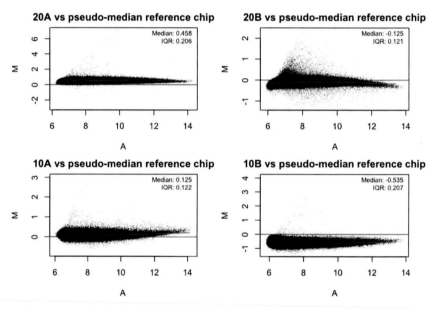

Fig. 14.5 MAplots for the four Dilution arrays before preprocessing

In addition to the MAplots for each of the four arrays, the MAplot function gives the median and IQR (interquartile) values of M. cex sets the font size of these data in the MAplots.

14.6.4 Background correction

affy has three methods for background correction: "none", mas, and rma. none does nothing, and is simply a placeholder. mas implements the background correction method used in the Affymetrix software. rma assumes that the PM probe intensities are the sum of a normal noise component with mean μ and variance σ^2, and a signal component that is exponential with mean α. The noise component is truncated at zero to avoid the possibility of negative intensity values. Given an observed intensity Y, the expectation value of the signal S is

$$E(S|Y=y) = a + b\frac{\phi(a/b) - \phi([y-a]/b)}{\Phi(a/b) + \Phi([y-a]/b) - 1} \qquad (14.3)$$

where $a = S - \mu - \sigma^2\alpha$, $b = \sigma$, ϕ is the normal density, and Φ is the normal distribution function. rma corrects only the PM probe intensities, while mas corrects both PM and MM intensities. Consult the Built-in processing methods vignette for details.

Applying rma and mas background correction methods leads to the following comparisons with the uncorrected data.

```
> Dilution.bg.rma = bg.correct(Dilution, method="rma")
> Dilution.bg.mas = bg.correct(Dilution, method="mas")
> par(mfrow=c(1,3))
> boxplot(Dilution, main="Dilution")
> boxplot(Dilution.bg.rma, main="Dilution.bg.rma")
> boxplot(Dilution.bg.mas, main="Dilution.bg.mas")
```

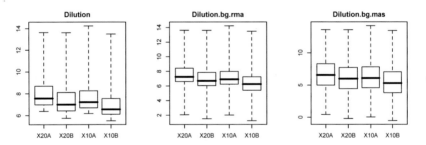

Fig. 14.6 Boxplots of four Dilution arrays after application of background correction methods: (left) none; (middle) rma; (right) mas

14.6.5 Normalization

affy has no fewer than seven normalization methods, whose names can be accessed by

```
> normalize.AffyBatch.methods
[1] "constant" "contrasts" "invariantset" "loess"
[5] "qspline" "quantiles" "quantiles.robust"
```

An explanation of each can be found in the `builtinMethods` vignette. The `quantiles` normalization method attempts to give each chip the same empirical distribution of probe intensities. The result of applying it to the background-corrected `Dilution` data is shown below, as the next step in harmonizing all four of the chips.

```
> Dil.bgrma.normq = normalize(Dilution.bg.rma,
  method="quantiles")
> par(mfrow=c(1,3))
> boxplot(Dilution, main="Dilution")
> boxplot(Dilution.bg.rma, main="Dilution.bg.rma")
> boxplot(Dil.bgrma.normq, main="Dil.bgrma.normq")
```

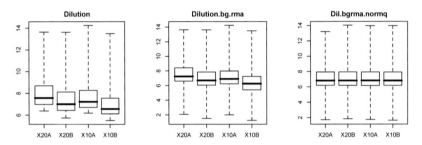

Fig. 14.7 Steps in harmonizing the four `Dilution` chips: (left) uncorrected; (middle) rma background correction; (right) quantiles normalization

14.6.6 PM correction

It may not be necessary to subtract mismatch probe intensities from perfect match intensities, but `affy` has three methods discussed in the `builtinMethods` vignette.

```
> pmcorrect.methods
[1] "mas" "pmonly" "subtractmm"
```

In `mas`, MM is subtracted from PM when MM < PM, and something else is subtracted otherwise to avoid negative numbers. `pmonly` will generally suffice.

14.6.7 Summarization

`affy` has five summarization methods discussed in the `builtinMethods` vignette.

```
> express.summary.stat.methods
[1] "avgdiff" "liwong" "mas" "medianpolish" "playerout"
```

Those used in the Affymetrix software ("mas") and in the Bioconductor RMA expression summary ("medianpolish") return results in log2 scale.

14.6.8 Combining preprocessing steps with *expresso*

Rather than doing each preprocessing step by itself, you can combine them with the affy function expresso. Various combinations can be used, though as the builtinMethods vignette points out, "It is important to note that not every pre-processing method can be combined together. In particular the rma backgrounds adjust only PM probe intensities and so they should only be used in conjunction with the pmonly PM correction. Also remember that the mas and medianpolish summarization methods $\log_2$ transform the data, thus they should not be used in conjunction with any preprocessing steps that are likely to yield negatives like the subtractmm pm correction method."

Here is the code to apply expresso to the Dilution data with a typical choice of methods, and the resulting output.

```
> eset = expresso(Dilution,bgcorrect.method="rma",
+ normalize.method="quantiles",
+ pmcorrect.method="pmonly",
+ summary.method="medianpolish")
background correction: rma
normalization: quantiles
PM/MM correction : pmonly
expression values: medianpolish
background correcting...done.
normalizing...done.
12625 ids to be processed
|                        |
|###################|
```

14.7 Using the results of preprocessing

14.7.1 ExpressionSet

Now that we have eset, what do we do with it? First of all, what kind of object is it? Asking for its class

```
> class(eset)
```

```
[1] "ExpressionSet"
attr(,"package")
[1] "Biobase"
```

tells us that it's an object of class ExpressionSet. The ExpressionSet help
page begins

```
Class to Contain and Describe High-Throughput
Expression Level Assays.

Description

Container for high-throughput assays and experimental
metadata. ExpressionSet class is derived from eSet,
and requires a matrix named exprs as assayData member.
```

The most important part of this description, for our present purposes, is that a matrix named exprs holds the processed probe intensity data. We can work with the
matrix as follows:

```
> Dil.expr = exprs(eset)
> summary(Dil.expr)
        20A                20B                10A                10B
  Min.   : 2.27    Min.   : 2.22    Min.   : 2.24    Min.   : 2.25
  1st Qu.: 4.72    1st Qu.: 4.69    1st Qu.: 4.73    1st Qu.: 4.70
  Median : 6.23    Median : 6.21    Median : 6.23    Median : 6.22
  Mean   : 6.19    Mean   : 6.18    Mean   : 6.19    Mean   : 6.18
  3rd Qu.: 7.52    3rd Qu.: 7.51    3rd Qu.: 7.51    3rd Qu.: 7.53
  Max.   :13.40    Max.   :13.40    Max.   :13.44    Max.   :13.49
> sd(Dil.expr)
   20A    20B    10A    10B
1.928  1.940  1.933  1.953
```

We see that the statistical properties of the four arrays are very similar as a result
of preprocessing. Remember that these results are expressed as $\log_2$ values, so a
mean of 6 corresponds to an expression value of $2^6 = 64$ and a standard deviation
of 2 corresponds to $2^2 = 4$ times the mean value.

14.7.2 Identifying highly expressed genes

We can get the expression values for the four arrays separately by writing

```
> e20A = Dil.expr[,1]
> e20B = Dil.expr[,2]
> e10A = Dil.expr[,3]
> e10B = Dil.expr[,4]
```

Then we can ask questions such as how many genes on array 20A have expressions
greater than three standard deviations above the mean:

```
> length(e20A[e20A > mean(e20A)+3*sd(e20A)])
[1] 55
```

To find the names of the genes that are most highly expressed, we combine feature names and expression values in a data frame, and sort it on expression value in decreasing order using the R order function (see R Help).

```
> e20A.df = data.frame(featureNames(eset),e20A)
> attach(e20A.df)
> e20A.order = order(e20A,featureNames.eset.,
  decreasing=T)
> # Row numbers of the 10 highest values:
> e20A.order[1:10]
 [1] 12625 10990  2819  5962   311  1974  1979  1261
      116  2488
> e20A.df[e20A.order[1:10],]
                 featureNames.eset.  X20A
AFFX-hum_alu_at     AFFX-hum_alu_at 13.40
40887_g_at              40887_g_at 13.38
32794_g_at              32794_g_at 13.21
35905_s_at              35905_s_at 13.07
1288_s_at                1288_s_at 13.04
31957_r_at              31957_r_at 12.93
31962_at                  31962_at 12.92
256_s_at                  256_s_at 12.77
1105_s_at                1105_s_at 12.69
32466_at                  32466_at 12.66
```

A Google search reveals that AFFX-hum_alu_at is an Affymetrix composite sequence, presumably added as a control.

14.8 Statistical analysis of differential gene expression

Once we have the adjusted and normalized gene expression levels, we want to ask various types of questions about the meaning of differential expression. Chapters 7–9 in the book by Stekel [60] are a good guide to this territory.

14.8.1 Paired samples

The following simulated data give the $\log_2$ expression data for the ACAT2 (acetyl-coenzyme A acetyltransferase 2) gene in 20 breast cancer patients before and after doxorubicin chemotherapy, relative to a common reference. We wish to determine

whether the ACAT2 gene is up- or down-regulated in breast cancer following treatment. The means and standard deviations in the calculation below come from Table
7.1 in [60], but the specific values are randomly drawn from a normal distribution
and will be different for each simulation.

```
> ACAT2.before = round(rnorm(20,-1.42,0.48),2)
> ACAT2.after = round(rnorm(20,-1.77,0.43),2)
> ACAT2.before
 [1] -1.57 -2.10 -1.52 -1.15 -1.34 -1.37 -1.78 -2.17
     -2.03 -0.69 -0.87 -0.70 -1.34
[14] -2.05 -1.34 -0.46 -1.15 -1.10 -1.03 -1.33
> ACAT2.after
 [1] -1.64 -1.79 -1.07 -1.68 -2.24 -2.20 -1.48 -2.42
     -1.60 -1.19 -1.94 -1.95 -1.80
[14] -1.46 -1.64 -2.04 -1.16 -1.56 -0.99 -1.94
> mean(ACAT2.before)
[1] -1.355
> mean(ACAT2.after)
[1] -1.690
```

Remember that a value of -1 means that the gene is down-regulated relative to the
standard by a factor of 2.

Since the samples are paired (the same patient before and after chemotherapy),
we use a paired t test (See Section 13.2).

```
> t.test(ACAT2.before,ACAT2.after,paired=T)

Paired t-test

data:  ACAT2.before and ACAT2.after
t = 2.492, df = 19, p-value = 0.02209
alternative hypothesis: true difference in means is
not equal to 0
95 percent confidence interval:
 0.05368 0.61632
sample estimates:
mean of the differences
                0.335
```

For this particular set of simulated values, the difference is significant at the 5%
level, but not at the 1% level. Since the data may not be normally distributed, we
can also use a Wilcoxon (Mann-Whitney) test.

```
> wilcox.test(ACAT2.before,ACAT2.after,paired=T)

Wilcoxon signed rank test with continuity correction

data:  ACAT2.before and ACAT2.after
```

```
V = 164.5, p-value = 0.02759
alternative hypothesis: true location shift is not
equal to 0

Warning message:
In wilcox.test.default(ACAT2.before, ACAT2.after,
paired = T) :
  cannot compute exact p-value with ties
```

The result is virtually the same.

14.8.2 Unpaired samples

Here we simulate data on the expression of the metallothionein IB gene in 27 patients with acute lymphoblastic leukemia (ALL) and 11 patients with acute myeloid leukemia (AML). Means and standard deviations come from Table 7.2 of [60].

```
> ALL = round(rnorm(27,7.93,0.94),2)
> AML = round(rnorm(11,8.97,0.51),2)
> ALL
 [1]  8.46  8.16  9.63  6.72  8.55  9.44  7.84  7.46
      8.84  7.67  8.64  5.67  6.93
[14]  8.06  7.82  7.40  8.15  6.40  8.14  6.75  9.27
      7.58  7.27 10.32  8.20  9.10
[27]  8.49
> AML
 [1] 9.25 9.38 8.64 9.30 9.97 9.02 8.97 9.20 8.46
     8.96 9.43
> t.test(ALL,AML,paired=F)

Welch Two Sample t-test

data:  ALL and AML
t = -4.681, df = 35.99, p-value = 3.969e-05
alternative hypothesis: true difference in means is
not equal to 0
95 percent confidence interval:
 -1.588 -0.628
sample estimates:
mean of x mean of y
    8.036     9.144
```

The difference in metallothionein IB expression levels is highly significant.

14.8.3 Bootstrapping

In small samples with potentially significant noise or experimental artifacts, the data may not be normally distributed and thus t tests would be inappropriate. In addition to the nonparametric Wilcoxon test, a method called *bootstrapping* is increasingly used. Bootstrapping proceeds by constructing a very large number (typically 1000 or more) of resamples of the observed dataset, and of equal size to the observed dataset, each of which is obtained by random sampling with replacement from the original dataset. We shall not consider bootstrapping here; see [60, Ch. 7] for some useful examples.

14.8.4 Analysis of variance (anova) for more than two groups

If more than two groups are being compared, we could either do multiple t tests, or we could do a test that compares all groups at once and returns a single *p*-value that gives the probability that one or more groups is different from the others. Suppose, for example, that we are measuring the expression of gene X in four groups of cancer patients: G1 (20 patients), G2 (37 patients), G3 (16 patients), and G4 (23 patients). We treat the group names as factors and simulate from normal distributions as follows:

```
> G = c(rep("G1", 20),rep("G2", 37), rep("G3", 16),
  rep("G4", 23))
> set.seed = 333 # So that simulation is reproducible
> G1 = rnorm(20,6.3,0.8)
> G2 = rnorm(37,7.0,0.6)
> G3 = rnorm(16,6.0,0.6)
> G4 = rnorm(23, 6.5, 0.8)
> # Concatenate the individual observations:
> Y = c(G1,G2,G3,G4)
```

We then compare two linear models: one in which Y does not depend on group membership (lm(Y ~ 1)) and one in which it does (lm(Y ~ G)). The anova function compares these two models and yields a probability that group membership is significant.

```
> anova(lm(Y ~ 1), lm(Y ~ G))
Analysis of Variance Table

Model 1: Y ~ 1
Model 2: Y ~ G
  Res.Df  RSS Df Sum of Sq      F  Pr(>F)
1     95 60.3
2     92 40.1  3      20.2 15.5 3.2e-08 ***
```

```
---
Signif. codes:  0 *** 0.001 ** 0.01 * 0.05 . 0.1   1
Warning message:
In model.matrix.default(mt, mf, contrasts) :
  variable 'G' converted to a factor
```

For this set of simulation parameters, group membership is highly significant with $p = 3.2 \times 10^{-8}$.

14.9 Detecting groups of genes

We are often interested in genes that are co-regulated, whose expression moves up or down together. There are a number of ways to explore this. To see some of the major approaches, we begin by constructing a simulated set of twelve genes whose expression will change with time in coordinated ways. All of the genes start at zero time near a $\log_2$ expression value of 6. The specific values are in the vector genestart. We then make four groups of three each: 1–3 increase with a slope of 0.5, so that (without random fluctuations) they would reach a $\log_2$ expression value of 12 after 12 hours; 4–6 grow 1/3 as fast, with a slope of 0.17; 7–9 do not change over time, on average; and 10–12 decrease with a slope of -0.25, so that they reach an average expression value of 3 after 12 hours. These slopes are in the vector geneslope. Finally, we add normal random fluctuations of mean zero and sd 0.5. The results over time are collected in the matrix genemat, which serves as the basis for further manipulation, and plotted with matplot.

```
genestart = c(6.0,6.1,5.9,6.2,5.8,6.3,5.7,6.0,6.1,5.9,
  6.2,5.8)
geneslope = c(0.5,0.5,0.5,0.17,0.17,0.17,0,0,0,-.25,
  -.25,-.25)
t = seq(0,12,by=2)
genemat = matrix(nrow=7,ncol=12) # Rows = time,
  columns = genes
for (i in 1:7) {
  for (j in 1:12) {
  genemat[i,j] = genestart[j]+(i-1)*2*geneslope[j] +
    rnorm(1,0,0.5)
  }}
> matplot(genemat,type="l",xlab="Time",
  main="Gene Expression")
> legend(1,13,legend=1:12,lty=1:12,col=1:12,bty="n")
```

We see that the expression behavior is more or less expected, though the random fluctuations somewhat obscure the differences between groups 4–6 and 7–9.

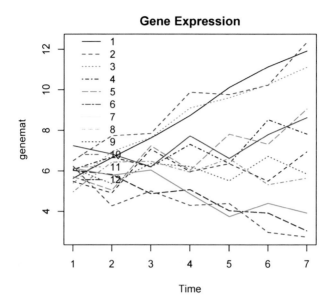

Fig. 14.8 Simulated behavior of three groups of genes with different expression behavior

14.9.1 Correlation

To see the correlations among the various groups, we use the `cor` function in R to calculate correlation functions. The correlation coefficient for each pair of genes will lie between 1 (perfect correlation) and −1 (perfect anticorrelation). A correlation coefficient near 0 indicates no correlation. There are three choices for the method in `cor`: `pearson`, `kendall`, and `spearman`. `pearson` is the standard method for calculating correlation coefficients, but it assumes a normal distribution and is susceptible to outliers. `spearman` is a nonparametric method and is robust to outliers, so it is a good choice for microarray analysis. However, it is insensitive to up-and-down patterns in time series, which could be a problem when profiling developmental patterns.

The results, shown in the top block on p. 299, are as expected. Try `matplot(C)` to see what it looks like.

14.9.2 Distance measures

The correlation matrix is a useful way to see which gene expression patterns are similar. However, even with a small set of twelve genes it is hard to discern patterns by inspection, and the task would become impossible with hundreds or thousands of genes. Therefore, statisticians have searched for ways to reduce the data in a multi-

```
> C = cor(genemat, method="spearman")
> round(C1,2)
       [,1]  [,2]  [,3]  [,4]  [,5]  [,6]  [,7]  [,8]  [,9] [,10] [,11] [,12]
 [1,]  1.00  0.93  0.93  0.93  0.96  0.82  0.64 -0.11  0.00 -0.75 -0.86 -0.75
 [2,]  0.93  1.00  1.00  0.89  0.96  0.89  0.39  0.00  0.00 -0.68 -0.79 -0.89
 [3,]  0.93  1.00  1.00  0.89  0.96  0.89  0.39  0.00  0.00 -0.68 -0.79 -0.89
 [4,]  0.93  0.89  0.89  1.00  0.86  0.79  0.50  0.14  0.00 -0.50 -0.68 -0.64
 [5,]  0.96  0.96  0.96  0.86  1.00  0.86  0.57 -0.07  0.04 -0.82 -0.89 -0.86
 [6,]  0.82  0.89  0.89  0.79  0.86  1.00  0.14  0.00  0.11 -0.61 -0.82 -0.71
 [7,]  0.64  0.39  0.39  0.50  0.57  0.14  1.00 -0.11  0.07 -0.68 -0.57 -0.29
 [8,] -0.11  0.00  0.00  0.14 -0.07  0.00 -0.11  1.00  0.57  0.14  0.07  0.21
 [9,]  0.00  0.00  0.00  0.00  0.04  0.11  0.07  0.57  1.00 -0.36 -0.32  0.25
[10,] -0.75 -0.68 -0.68 -0.50 -0.82 -0.61 -0.68  0.14 -0.36  1.00  0.93  0.61
[11,] -0.86 -0.79 -0.79 -0.68 -0.89 -0.82 -0.57  0.07 -0.32  0.93  1.00  0.61
[12,] -0.75 -0.89 -0.89 -0.64 -0.86 -0.71 -0.29  0.21  0.25  0.61  0.61  1.00
```

```
> genedist = 1-C
> round(genedist,2)
       [,1]  [,2]  [,3]  [,4]  [,5]  [,6]  [,7]  [,8]  [,9] [,10] [,11] [,12]
 [1,]  0.00  0.07  0.07  0.07  0.04  0.18  0.36  1.11  1.00  1.75  1.86  1.75
 [2,]  0.07  0.00  0.00  0.11  0.04  0.11  0.61  1.00  1.00  1.68  1.79  1.89
 [3,]  0.07  0.00  0.00  0.11  0.04  0.11  0.61  1.00  1.00  1.68  1.79  1.89
 [4,]  0.07  0.11  0.11  0.00  0.14  0.21  0.50  0.86  1.00  1.50  1.68  1.64
 [5,]  0.04  0.04  0.04  0.14  0.00  0.14  0.43  1.07  0.96  1.82  1.89  1.86
 [6,]  0.18  0.11  0.11  0.21  0.14  0.00  0.86  1.00  0.89  1.61  1.82  1.71
 [7,]  0.36  0.61  0.61  0.50  0.43  0.86  0.00  1.11  0.93  1.68  1.57  1.29
 [8,]  1.11  1.00  1.00  0.86  1.07  1.00  1.11  0.00  0.43  0.86  0.93  0.79
 [9,]  1.00  1.00  1.00  1.00  0.96  0.89  0.93  0.43  0.00  1.36  1.32  0.75
[10,]  1.75  1.68  1.68  1.50  1.82  1.61  1.68  0.86  1.36  0.00  0.07  0.39
[11,]  1.86  1.79  1.79  1.68  1.89  1.82  1.57  0.93  1.32  0.07  0.00  0.39
[12,]  1.75  1.89  1.89  1.64  1.86  1.71  1.29  0.79  0.75  0.39  0.39  0.00
```

dimensional "space" to a few synoptic variables. The essential idea is that genes that are expressed similarly to each other will be "close" to each other in some suitably defined space; the "distance" between them in that space will be small.

If two gene expression patterns are identical, their correlation coefficient will be 1, and the distance between them will be 0. Therefore, one commonly used measure of distance between any two samples i and j is $1 - C_{ij}$, where C is the correlation matrix. Applying this to our simulated dataset we get the gene distances shown in the lower block on p. 299. Both the spearman and pearson methods are commonly used to determine distances. Another distance measure between two expression profiles X and Y is the "Euclidian distance", defined as

$$d(X,Y) = \left[\sum_{i=1^n} (x_i - y_i)^2 \right]^{1/2} \qquad (14.4)$$

where x_i and y_i are the results of the ith measurement. Stekel [60, Ch. 8] has a readable discussion of the relative virtues of these approaches. In what follows we will use the Spearman distances we have already calculated.

14.9.3 Clustering

Having obtained the distances between gene profiles, we want to ask which profiles belong together. We shall look at two ways for doing this: clustering and principal component analysis. (See [23, Chs. 13 and 15] for readable overviews of these topics.) Clustering, which is probably the most commonly used method, attempts to partition a dataset into subsets whose elements share common features. A useful survey of clustering analysis of genomic data is Chapter 13 by Pollard and van der Laan in the book by Gentleman et al. [27].We apply two of the clustering methods demonstrated in that chapter to our genedist data.

We first load the cluster library, then apply the pam (Partitioning Around Medoids) function, using k=4 clusters (since that's how we set the simulation up). According to Wikipedia, "Medoids are representative objects of a data set. ... Basically, the algorithm finds the center of a cluster and takes the element closest to the center as "the medoid" after which, the distances to the other points are computed."

```
> library(cluster)

> pam(genedist,k=4)
Medoids:
      ID
[1,]   3 0.07143 2.220e-16 0.0000 0.1071 0.03571 0.1071
         0.6071 1.0000 1.0000 1.67857
[2,]   7 0.35714 6.071e-01 0.6071 0.5000 0.42857 0.8571
         0.0000 1.1071 0.9286 1.67857
```

```
[3,]  9 1.00000 1.000e+00 1.0000 1.0000 0.96429 0.8929
         0.9286 0.4286 0.0000 1.35714
[4,] 11 1.85714 1.786e+00 1.7857 1.6786 1.89286 1.8214
         1.5714 0.9286 1.3214 0.07143

[1,] 1.786 1.8929
[2,] 1.571 1.2857
[3,] 1.321 0.7500
[4,] 0.000 0.3929
Clustering vector:
 [1] 1 1 1 1 1 1 2 3 3 4 4 4
Objective function:
 build   swap
0.3958 0.3129

Available components:
 [1] "medoids"    "id.med"     "clustering" "objective"
     "isolation"  "clusinfo"
 [7] "silinfo"    "diss"       "call"       "data"
```

Ignoring most of this output, we see that the clustering vector has put genes 1–6 in cluster 1, gene 7 into a cluster of its own, genes 8 and 9 into cluster 3, and genes 10–12 into cluster 4. In terms of our starting model, all of the increasing-expression genes were put in one cluster, as were the decreasing-expression genes; but the "unchanging" genes were subdivided. The clusters can be viewed graphically by the following code.

```
> part = pam(genedist,k=4)
> plot(part,which.plots=1,labels=3,main="PAM")
```

A different clustering algorithm, which yields a different graphical representation, is diana (DIvisive ANAlysis Clustering).

```
> hier = diana(genedist)
> plot(hier, which.plots=2, main="DIANA")
```

This hierarchical clustering diagram shows two major classes, corresponding roughly to increasing and decreasing gene expression, but subdivides the classes somewhat differently than pam. Comparison of these two methods shows that clustering analysis of microarrays is not cut-and-dried, and that different methods should be thoughtfully compared. Read R Help for pam and diana to learn more about how they work.

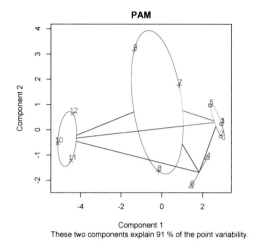

Fig. 14.9 Clustering of the three groups of genes using the PAM method

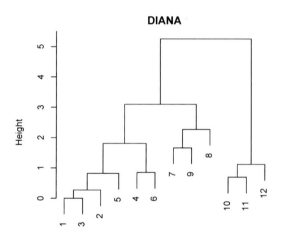

genedist
Divisive Coefficient = 0.83

Fig. 14.10 Clustering of the three groups of genes using the DIANA method

14.9.4 Principal component analysis

Principal component analysis (PCA) transforms and reduces multidimensional datasets to lower dimensions for analysis. When the multivariate dataset is visualized as a set of coordinates in a high-dimensional data space (one axis per variable), PCA gives a two-dimensional projection viewed from the most informative viewpoint, that which maximizes the variance of the data. This 2D image has the first and second principal components as axes, with samples or metadata (e.g., names of factors describing groups of samples) as points on the 2D surface. A detailed discussion of principal component analysis using R is given in [64, pp. 302–313]. The function to call is `prcomp`, and the results are shown in the top block on p. 304.

The `Rotation` matrix tells how the axes of the original 12-dimensional vector must be rotated to give the principal components. It is not of direct interest to us here. The `summary` method applied to `procomp` gives the cumulative proportion for each principal component, with results shown in the bottom block on p. 304. We see that 99% of the variance is in the first three principal components. The relative contributions can be visualized in a bar plot.

```
> plot(prcomp(genedist)
```

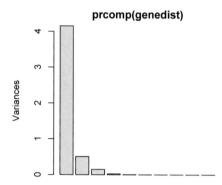

Fig. 14.11 Relative weights of the principal components for the three groups of genes

The `biplot` function shows the projects of the 12 gene expression variables on the two principal component axes.

```
biplot(prcomp(genedist))
```

Groupings emerge that are similar, but not identical, to those obtained from cluster analysis. See Fig. 14.12.

14.10 Power analysis

In planning and evaluating microarray experiments, we want to know their statistical significance. We have already discussed t tests and their nonparametric Wilcoxon counterparts, which give the probability of avoiding Type I errors (accepting a false positive hypothesis.) We also need to know the power of an experiment in avoiding Type II errors, i.e., rejecting a false negative hypothesis. This is the realm of power analysis, which we have introduced in the previous chapter, Section 13.2.5. The

```
> prcomp(genemat)
Standard deviations:
[1] 4.266e+00 1.049e+00 7.187e-01 6.145e-01 3.875e-01 2.903e-01 5.484e-16

Rotation:
             PC1       PC2      PC3       PC4       PC5       PC6       PC7
[1,]    0.497217   0.08726 -0.59241  0.199605 -0.004361 -0.184910  0.334031
[2,]    0.543867  -0.09815  0.52007  0.113151  0.001247 -0.291914  0.133740
[3,]    0.427988   0.37629  0.13975  0.106063  0.099224 -0.049414 -0.661729
[4,]    0.129796   0.23954  0.14212  0.166978  0.046830  0.500592  0.290590
[5,]    0.203160  -0.14309  0.05229 -0.204409  0.298627  0.541565 -0.036816
[6,]    0.154132  -0.09286 -0.21631 -0.170377  0.699208 -0.006608 -0.095934
[7,]    0.011820  -0.04221 -0.03211 -0.002709 -0.368264 -0.023602 -0.404161
[8,]   -0.012284  -0.07724  0.48221  0.235061  0.065474  0.121445  0.234988
[9,]   -0.006502  -0.33316  0.05085  0.255885  0.141011 -0.388390  0.116917
[10,]  -0.207478   0.44969 -0.10619  0.669095  0.114309  0.088352 -0.006855
[11,]  -0.250635   0.56663  0.20229 -0.358312  0.304085 -0.393262  0.179015
[12,]  -0.292805  -0.33563  0.03140  0.374966  0.380194 -0.074904 -0.270781

> summary(prcomp(genedist))
Importance of components:
                         PC1    PC2    PC3     PC4     PC5     PC6     PC7     PC8
Standard deviation     2.191 0.4903 0.3487 0.17989 0.09420 0.02965 0.00457 0.00185
Proportion of Variance 0.922 0.0462 0.0234 0.00622 0.00171 0.00017 0.00000 0.00000
Cumulative Proportion  0.922 0.9686 0.9919 0.99812 0.99983 1.00000 1.00000 1.00000
                            PC9     PC10     PC11     PC12
Standard deviation     0.000783 0.000359 7.27e-17 6.07e-18
Proportion of Variance 0.000000 0.000000 0.00e+00 0.00e+00
Cumulative Proportion  1.000000 1.000000 1.00e+00 1.00e+00
```

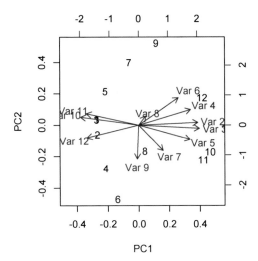

Fig. 14.12 Biplot representation of principal component analysis

practical question is, how many replicates do we need to achieve a desired power? Stekel [60, pp. 221–228] has a lucid discussion of these issues; our examples are motivated by his. R Help for `power.t.test` tells us that the arguments of the function are

```
power.t.test(n = NULL, delta = NULL, sd = 1,
   sig.level = 0.05, power = NULL,
      type = c("two.sample", "one.sample", "paired"),
      alternative = c("two.sided", "one.sided"),
      strict = FALSE)
Arguments
```

```
n   Number of observations (per group)
delta   True difference in means
sd   Standard deviation
sig.level   Significance level (Type I error probability)
power   Power of test (1 minus Type II error probability)
type   Type of t test
alternative   One- or two-sided test
strict   Use strict interpretation in two-sided case
```

Before proceeding further, we note that if, as is usually assumed, the distribution of errors in a microarray experiment follows a log-normal distribution, then the variance σ^2 of the log of the signal is related to the coefficient of variability v by ([60, p. 216])

$$\sigma^2 = \ln(v^2 + 1) \tag{14.5}$$

Since this equation uses natural logs, and microarray expression data are $\log_2$, we must divide the calculated standard deviation σ by $\ln(2) = 0.693$ to get the standard deviation in base 2. Expressing this as a function in R

```
sd.2 = function(v) sqrt(log(v^2+1))/log(2)
```

For example, if the coefficient of variation is 0.5,

```
sd.2(.5)
[1] 0.6815
```

As a first example, suppose we are conducting a study of 22 cancer patients treated with a new drug. We run microarrays on samples taken before and after treatment, and analyze for up- and down-regulated genes using a t test. We will be analyzing 7000 genes, and want no more than one false positive, so sig.level = 1/7000. The coefficient of variability is 0.5, leading to a standard deviation (base 2) of 0.68. Since the same patients are studied before and after, this is a one-sample test. We want to know the power of detecting 2-fold regulated genes (either up or down). On a $\log_2$ basis, this corresponds to a delta of 1. The power of this experiment is then calculated to be

```
> power.t.test(n=22,delta=1,sd=0.68,sig.level=1/7000,
  type="one.sample",alt="two.sided")

    One-sample t test power calculation

              n = 22
          delta = 1
             sd = 0.68
      sig.level = 0.0001429
          power = 0.9697
    alternative = two.sided
```

That is, if our significance level threshold is high enough to have just one false positive in 7000 genes, our analysis should yield 97% of the genes whose expression is modified twofold.

To determine what degree of expression can be detected at the 99% power level, we put in 0.99 for power and leave out delta, which is returned by the calculation.

```
> power.t.test(n=22,power=0.99,sd=0.68,
  sig.level=1/7000,type="one.sample",alt="two.sided")

    One-sample t test power calculation

              n = 22
          delta = 1.082
             sd = 0.68
      sig.level = 0.0001429
          power = 0.99
    alternative = two.sided
```

This corresponds to a $2^{1.082} = 2.117$-fold expression change.

Finally, suppose that we were testing another drug, and were going to look at 12,000 genes with only one false positive. How many patients would we need to get 99% power with `delta = 1`?

```
> power.t.test(power=0.99,delta=1,sd=0.68,
   sig.level=1/12000,type="one.sample", alt="two.sided")

       One-sample t test power calculation

              n = 25.77
          delta = 1
             sd = 0.68
      sig.level = 8.333e-05
          power = 0.99
    alternative = two.sided
```

Rounding up, we would need 26 patients in the study.

14.11 Problems

1. Get the sample names, the first ten gene names, and the first ten probe names of the `Dilution` dataset. Which probe names correspond to `"1000_at"`? Which gene names correspond to `"1000_at"`? How many probe names are in `Dilution`? How many genes?

2. Get the PM and MM intensities of the first ten probes in `Dilution`, and subtract the mismatch from the perfect match intensities. Are there any negative values? What does this imply about using mismatch probes to determine the baseline?

3. Make MAplots of the dilution data after background adjustment by the `rma` and `mas` methods. Do you see any obvious differences between the two methods?

4. Normalize the `Dilution.bg.rma` data with the `constant`, `loess`, and `invariantset` methods, and plot the data as side-by-side, suitably labeled boxplots along with the `quantiles` normalization. Comment on any obvious differences you observe.

5. Make MAplots of the four baseline-corrected and normalized datasets from the previous problem. Use `par(mfrow=c(2,2))` to arrange the plots in a 2×2 matrix. Are the MAplots now essentially parallel to the A axis and normally scattered around $M = 0$?

6. Apply `expresso` to the `Dilution` data, using the standard Affymetrix MAS choices of methods. Compare the summaries and standard deviations for the four chips with those obtained by the methods in the text.

7. Plot histograms of the expression data for the four `Dilution` samples, and overlay normal distribution curves with the same mean and standard deviation.

Also do qqnorm plots of the four expression sets. Do the data appear to be normally distributed?

8. Generate new simulated data and redo the calculations in Section 14.9 using both spearman and pearson correlation methods on the same data. Are there any notable differences?

9. Calculate how many patients would be needed in the last power analysis example if there were two groups of patients, one treated and one untreated, so that a "two.sample" power.t.test was required.

10. Find a recent journal article that deals with bootstrap resampling of microarray data, and summarize its findings.

Appendix A
Basic String Manipulations in R

In this Appendix we summarize some of the basic string manipulation capabilities of R. For more powerful and complex capabilities, especially regular expressions, consult Chapter 5 in Gentleman [26].

Elementary string manipulations

First, use a simple test string to show the substring function.

```
> ts="abcde"
> substring(ts,1,1) # What is the first character in ts?
[1] "a"
> substring(ts,1,1)="X" # Change the first character to "X"
> ts
[1] "Xbcde"
> substring(ts,1,3) # What are the first three characters in ts?
[1] "Xbc"
```

nchar counts the number of characters in a string, which is different from the length (number of elements) of the string vector.

```
> nchar(ts) # Number of characters in ts
[1] 5
> length(ts) # Length of the vector ts
[1] 1
```

As a more relevant biological example, we examine the beginning of the lambda bacteriophage genome. We start with the first 70 bases.

```
lam70="GGGCGGCGACCTCGCGGGTTTTCGCTATTTATGAAAATTTTCCGGTT
       TAAGGCGTTTCCGTTCTTCTTCG"
> nchar(lam70)
[1] 70
# Keep the first 60 bases in lam60
```

```
> lam60 = substr(lam70,1,60)
> lam60
[1] "GGGCGGCGACCTCGCGGGTTTTCGCTATTTATGAAAATTTTCCGG
    TTTAAGGCGTTTCCG"
> nchar(lam60)
[1] 60
```

Now we break lam60 into 20 triplets with substring.

```
> lam60.triplets = substring(lam60, seq(1,58,3),
  seq(3,60,3))
> lam60.triplets
 [1] "GGG" "CGG" "CGA" "CCT" "CGC" "GGG" "TTT"
     "TCG" "CTA" "TTT" "ATG" "AAA" "ATT" "TTC"
[15] "CGG" "TTT" "AAG" "GCG" "TTT" "CCG"
```

The related function substr would yield just the first triplet.
The reverse of substring is paste.

```
paste(lam60.triplets,sep="")
> paste(lam60.triplets,sep="", collapse="")
[1] "GGGCGGCGACCTCGCGGGTTTTCGCTATTTATGAAAATTTTCCGGTTT
    AAGGCGTTTCCG"
```

The R base installation has functions toupper and tolower to convert strings
between upper and lower case. The function chartr translates specified characters, and is useful to convert from DNA to RNA.

```
> tolower("ABCDE")
[1] "abcde"
> toupper("abcde")
[1] "ABCDE"
> chartr("T","U",lam60)
[1] "GGGCGGCGACCUCGCGGGUUUUCGCUAUUUAUGAAAAUUUUCCGGUUU
    AAGGCGUUUCCG"
```

chartr can change more than one character at a time.

```
> chartr("ab","12","aabb")
[1] "1122"
```

We can thus use chartr to obtain the complement of the sequence.

```
> complam60 = chartr("ACGT", "TGCA", lam60)
> paste(rev(unlist(strsplit(complam60,split=""))),
  collapse="")
[1] "CGGAAACGCCTTAAACCGGAAAATTTTCATAAATAGCGAAAACCCGC
    GAGGTCGCCGCCC"
```

By combining chartr to convert the bases to their complements, strsplit to
split the complement string into its components (remember to convert the resulting

list into a vector with `unlist`), `rev` to reverse the elements of the vector, and `paste` to recombine the vector elements into a string, one can write a function `revcomp` to obtain the reverse complement of the original DNA string.

```
> revcomp = function(DNAstr) {
step1 = chartr("ACGT","TGCA",DNAstr)
step2 = unlist(strsplit(step1, split=""))
step3 = rev(step2)
step4 = paste(step3, collapse="")
return(step4)
}
> revcomp(lam60)
[1] "CGGAAACGCCTTAAACCGGAAAATTTTCATAAATAGCGAAAACCCGC
    GAGGTCGCCGCCC"
```

Genetic code translation

The Bioconductor package `Biostrings` contains the genetic code to convert triplets of bases to amino acids.

```
> source("http://bioconductor.org/biocLite.R")
biocLite("Biostrings")

> library(Biostrings)

Attaching package: 'Biostrings'

> paste(GENETIC_CODE[lam60.triplets],collapse="")
[1] "GRRPRGFSLFMKIFRFKAFP"
> strsplit("GRRPRGFSLFMKIFRFKAFP",split="")
[[1]]
 [1] "G" "R" "R" "P" "R" "G" "F" "S" "L" "F" "M" "K"
     "I" "F" "R" "F" "K" "A" "F" "P"
```

`Biostrings` contains the method `AMINO_ACID_CODE` to translate one-letter to three-letter amino acid codes. It selects the array element corresponding to the one-letter code, and replaces it with the three-letter equivalent. We apply it to the amino acid sequence above.

```
> AMINO_ACID_CODE[strsplit(as.character("GRRPRGFSLFMK
  IFRFKAFP"), NULL)[[1]]]
    G     R     R     P     R     G     F     S     L
"Gly" "Arg" "Arg" "Pro" "Arg" "Gly" "Phe" "Ser" "Leu"
    F     M     K     I     F
"Phe" "Met" "Lys" "Ile" "Phe"
```

```
    R      F      K      A      F      P
  "Arg"  "Phe"  "Lys"  "Ala"  "Phe"  "Pro"
```

Two things should be noted about this code. `strspit` returns a list, and `[[1]]` picks out the first (and only) element of the list, which is a vector. The outer square brackets then denote the element of the vector to which `AMINO_ACID_CODE` is applied.

`Biostrings` also has functions to reverse or complement nucleic acid sequences: `reverse`, `complement`, and `reverseComplement` operate on DNA and RNA sequences, which must be declared as such. See `help(reverse)` for details.

References

1. Adler, F.: Modeling the Dynamics of Life: Calculus and Probability for Life Scientists, 2nd edn. Brooks/Cole (2005)
2. Allman, E., Rhodes, J.: Mathematical Models in Biology: An Introduction. Cambridge University Press (2004)
3. Alon, U.: An Introduction to Systems Biology: Design Principles of Biological Circuits. Chapman & Hall/CRC, Boca Raton, FL (2007)
4. Altman, D.: Practical Statistics for Medical Research. Chapman & Hall (1991)
5. Barkai, N., Leibler, S.: Robustness in simple biochemical networks. Nature **387**, 913–917 (1997)
6. Beard, D., Qian, H.: Chemical Biophysics: Quantitative Analysis of Cellular Systems. Cambridge University Press (2008)
7. Berg, H.: E. coli in Motion. Springer (2003)
8. Bloomfield, V., Crothers, D., Tinoco Jr., I.: Nucleic Acids: Structures, Properties, and Functions. University Science Books (2000)
9. Boroujerdi, M.: Pharmacokinetics: Principles and Applications. McGraw-Hill (2002)
10. Braun, M.: Differential Equations and Their Applications. An Introduction to Applied Mathematics, 4th edn. Springer (1993)
11. Cicirelli, M., Smith, L.: Cyclic amp levels during the maturation of xenopus oocytes. Dev. Biol. **108**, 254–258 (1985)
12. Council, N.R.: BIO 2010: Transforming Undergraduate Education for Future Research Biologists. The National Academies Press (2003)
13. Crank, J.: The Mathematics of Diffusion, 2nd edn. Oxford University Press (1975)
14. Cropper, W.: Mathematica Computer Programs for Physical Chemistry. Springer (1998)
15. Dalgaard, P.: Introductory Statistics with R. Springer (2002)
16. von Dassow, G., Meir, E., Munro, E., Odell, G.: The segment polarity network is a robust developmental module. Nature **406**, 188–192 (2000)
17. von Dassow, G., Odell, G.: Design and constraints of the *Drosophila* segment polarity module: Robust spatial patterning emerges from intertwined cell state switches. J. Exp. Zool. Mol. Dev. Evol. **294**, 179–215 (2002)
18. Deonier, R., Tavaré, S., Waterman, M.: Computational Genome Analysis: An Introduction. Springer (2005)
19. Diekmann, O., Heesterbeek, J.: Mathematical Epidemiology of Infectious Diseases: Model Building, Analysis and Interpretation. John Wiley & Sons (2000)
20. Eldar, A., Dorfman, R., Weiss, D., Ashe, H., Shilo, B.Z., Barkai, N.: Robustness of the bmp morphogen gradient in *Drosophila* embryonic patterning. Nature **419**, 304–308 (2002)
21. Eldar, A., Rosin, D., Shilo, B.Z., Barkai, N.: Self-enhanced ligand degradation underlies robustness of morphogen gradients. Dev. Cell **5**, 635–646 (2003)

22. Eldar, A., Shilo, B.Z., Barkai, N.: Elucidating mechanisms underlying robustness of morphogen gradients. Curr. Opin. Genet. Dev. **14**, 436–439 (2004)
23. Everitt, B., Hothorn, T.: A Handbook of Statistical Analyses Using R. Chapman & Hall/CRC (2006)
24. Fall, C., Marland, E., Wagner, J., Tyson, J. (eds.): Computational Cell Biology. Springer (2002)
25. Fell, D.: Understanding the Control of Metabolism. Portland Press (1997)
26. Gentleman, R.: R Programming for Bioinformatics. Chapman & Hall/CRC (2008)
27. Gentleman, R., Carey, V., Huber, W., Irizarry, R., Dudoit, S. (eds.): Bioinformatics and Computational Biology Solutions Using R and Bioconductor. Springer (2005)
28. Gillespie, D.: Exact stochastic simulation of coupled chemical reactions. J. Phys. Chem **81**, 2340 (1977)
29. Gillespie, D.: Stochastic simulation of chemical kinetics. Annu. Rev. Phys. Chem. **58**, 35 (2007)
30. Groen, A., Westerhoff, H.: Modern control theories: A consumers' test. In: A. Cornish-Bowden, M. Cardenas (eds.) Control of Metabolic Processes, chap. 6, pp. 101–118. Plenum Press (1990)
31. Hahne, F., Huber, W., Gentleman, R., Falcon, S.: Bioconductor Case Studies. Springer (2008)
32. Heinrich, R., Rapoport, T.: A linear steady-state treatment of enzymatic chains: general properties, control and effector strength. Eur. J. Biochem. **42**, 89–95 (1974)
33. Heinrich, R., Schuster, S.: The Regulation of Cellular Systems. Chapman & Hall (1996)
34. Holling, C.: The components of predation as revealed by a study of small mammal predation of the european pine sawfly. Can. Entomol. **91**, 293–320 (1959)
35. Ingolia, N.: Topology and robustness in the *Drosophila* segment polarity network. PLoS Biol. **2**(6), 805–815 (2004)
36. Jones, O., Maillardet, R., Robinson, A.: Introduction to Scientific Programming and Simulation Using R. Chapman & Hall/CRC (2009)
37. Kacser, H., Burns, J.: The control of flux. Symp. Soc. Exp. Biol. **27**, 65–104 (1973)
38. Keeling, M., Rohani, P.: Modeling Infectious Diseases in Humans and Animals. Princeton University Press (2008)
39. Keen, R., Spain, J.: Computer Simulation in Biology: A BASIC Introduction. Wiley-Liss (1992)
40. Kermack, W., McKendrick, A.: A contribution to the mathematical theory of epidemics. Proc. R. Soc. London, Ser. A **115**, 700–721 (1927)
41. Leslie, P., Gower, J.: Properties of a stochastic model for the predator-prey type of interaction between two species. Biometrika **47**, 219–301 (1960)
42. Macheras, P., Iliadis, A.: Modeling in Biopharmaceutics, Pharmacokinetics, and Pharmacodynamics: Homogeneous and Heterogeneous Approaches. Springer (2006)
43. Maindonald, J., Braun, J.: Data Analysis and Graphics Using R: An Example-Based Approach, 2nd edn. Cambridge University Press (2007)
44. Mangan, S., Alon, U.: Structure and function of the feed-forward loop network motif. PNAS **100**(21), 11,980–11,985 (2003)
45. Milanese, M., Molino, G.: Structural identifiability of compartmental models and pathophysiological information from the kinetics of drugs. Math. Biosci. **26**, 175–190 (1975)
46. Molino, G., Milanese, M.: Structured analysis of compartmental models for the hepatic kinetics of drugs. J. Lab. Clin. Med. **85**, 865–878 (1975)
47. Murrell, P.: R Graphics. Chapman & Hall/CRC (2006)
48. Nowak, M., May, R.: Virus Dynamics: Mathematical Principles of Immunology and Virology. Oxford University Press (2000)
49. Paradis, E.: Analysis of Phylogenetics and Evolution with R. Springer (2006)
50. Pearl, R.: The growth of populations. Q. Rev. Biol. **2**, 532–548 (1927)
51. Powers, W., Canale, R.: Some applications of optimization techniques to water quality modeling and control. IEEE Trans. Syst., Man, Cybernetics **5**, 312–321 (1975)
52. Samuels, M., Witmer, J.: Statistics for the Life Sciences, 3rd edn. Pearson Education (2003)

53. SantaLucia Jr., J., Allawi, H., Seneviratne, P.: Improved nearest-neighbor parameters for predicting dna duplex stability. Biochemistry **35**, 3555–3562 (1996)
54. Savageau, M.: Biochemical Systems Analysis: A Study of Function and Design in Molecular Biology. Addison-Wesley (1976)
55. Seefeld, K., Linder, E.: Statistics Using R with Biological Examples. http://cran.r-project.org/doc/contrib/Seefeld_StatsRBio.pdf (2007)
56. Simon, R., Korn, E., McShane, L., Radmacher, M., Wright, G., Zhao, Y.: Design and Analysis of DNA Microarray Investigations. Springer (2003)
57. Smith, H., Waltman, P.: The Theory of the Chemostat: Dynamics of Microbial Competition. Cambridge University Press (2008)
58. Staub, J.F., Brezillon, P., Perault-Staub, A., Milhaud, G.: Nonlinear modeling of calcium metabolism—first attempt in the calcium-deficient rat. Trans. Inst. Measur. Control (London) **3**, 89–97 (1981)
59. Steinmetz, C., Larter, R.: The quasiperiodic route to chaos in a model of the peroxidase-oxidase reaction. J. Chem. Phys. **94**(2), 1388–1396 (1991)
60. Stekel, D.: Microarray Bioinformatics. Cambridge University Press (2003)
61. Tanner, J.: The stability and intrinsic growth rates of predator and prey populations. Ecology **56**, 855–867 (1975)
62. Tinoco Jr., I., Sauer, K., Wang, J., Puglisi, J.: Physical Chemistry: Principles and Applications in Biological Sciences, 4th edn. Prentice-Hall (2002)
63. Torres, N., Voit, E.: Pathway Analysis and Optimization in Metabolic Engineering. Cambridge University Press (2002)
64. Venables, W., Ripley, B.: Modern Applied Statistics with S, 4th edn. Springer (2002)
65. Verzani, J.: Using R for Introductory Statistics. Chapman & Hall/CRC (2005)
66. Voit, E.: Computational Analysis of Biochemical Systems: A Practical Guide for Biochemists and Molecular Biologists. Cambridge University Press (2000)
67. Wilkinson, D.: Stochastic Modelling for Systems Biology. Chapman & Hall/CRC (2006)
68. Wolpert, L.: Positional information and the spatial pattern of cellular differentiation. J. Theor. Biol. **25**, 1–47 (1969)
69. Xia, T., SantaLucia Jr., J., Burkard, M., Kierzek, R., Schroeder, S., Jiao, X., Cox, C., Turner, D.: Thermodynamic parameters for an expanded nearest-neighbor model for formation of rna duplexes with watson-crick base pairs. Biochemistry **37**, 14,719–14,735 (1998)

Index

Printed by Books on Demand, Germany

: